SCIENCE, TECHNOLOGY AND HUMAN VALUES

S Ojha

MD PUBLICATIONS PVT LTD
NEW DELHI
www.mdppl.com

Published by :

MD Publications Pvt Ltd
"MD House", 11, Darya Ganj,
New Delhi - 110 002
Phone : +91-11- 45355555
E-mail : contact@mdppl.com
Website: www.mdppl.com

ISBN : 978-81-7533-326-0

Published and Printed by Mr. Pranav Gupta on behalf of **MD Publications Pvt Ltd** at Repro India Limited.

SCIENCE, TECHNOLOGY AND HUMAN VALUES

PREFACE

Science, Technology and Human Values examine normative issues in the conduct of science and the development and implementation of technology. Proposals focus on how ethical issues and values interconnect with science and technology, and how norms and values institutionalized in science and technology engage with society.

The advent and progress of technology as it becomes a global all-encompassing phenomenon appears to be both a blessing in the way it improves the human condition in its efforts at coping with life and the world and as a tragedy in its destructive outcome as it is affecting the physical universe and impacting on the future of humanity. It lessens the hardship, the suffering, and the despair in the face of overwhelming odds in mastering the universe and achieving chosen goals and enhances self-mastery, higher achievements and hope in performing essential tasks and chores demanded or chosen for survival or human flourishing.

Technology as the ever-present dynamic factor affecting our daily lives and transforming our contemporary civilization could be described as the rationally created artificial world of learnable operational rules, methods, recipes, and skills furnishing a complex of universally available standardized means used automatically to attain predetermined goals in any variety of endeavors with more mastery, more efficiency, more productivity, more predictability, more durability, and more practicality. Technology, under all its forms, dominates every aspect of our lives by making it possible for us to manipulate any physical or mental activity dealing with domestic, social, political, economic, medical and aesthetic concerns, facilitating means of transportation and communication, enhancing conditions of work and play, and helping in harnessing the forces of nature and in transforming raw material.

As scientific advances improve our lives, they also complicate how we live and react to the new technologies. More and more, changes in human values intersect knowledge production practices, and come into conflict with scientific advancement as we deal with important issues such as nuclear power, environmental degradation, and information technology. The implementation of technology influences the values of a society by changing expectations and realities. The implementation of technology is also influenced by values.

I am very grateful to Mr. Pranav Gupta, Director of MD Publications Pvt Ltd, New Delhi, for his kind cooperation and support for this book.

S Ojha

Contents

1

Science, Technology and Human Values

An Introduction

Ideally, new technology always advances, enhances and supports human values. But of course this is not an ideal world. The effects of technology are mixed. For example, the "agricultural revolution" and the "industrial revolution" brought many benefits to human beings: food for the hungry, effective medical care for the sick, relief from heavy labor, rapid and comfortable transportation, and so on. Nevertheless, problems were generated: overpopulation, world threatening weapons, pollution, terrible accidents which killed many people, etc. Too often, new technology develops with little attention to its impact upon human values.

The mass production of automobiles, for example, had profound effects upon cities, travel, entertainment, nature, the environment, even sexual mores. Many of the consequences were unforeseen – even unimagined – by those who created the technology. Let us do better! In particular, let us do what we can in this era of "the computer revolution" to see that computer technology advances human values. True enough, we could argue endlessly over the meanings of terms like "privacy," "health," "security," "fairness," or "ownership." Philosophers do it all the time – and ought to. But people understand such values well enough to desire and even to treasure them. We do not need absolute clarity or unattainable unanimity before we do anything to advance them.

Let us look at human values as a scientist looks at the phenomenon he chose for studying. What do we mean by human values? First of all, this is something we appreciate and want to have, or achieve. Values are something we qualify as good, and are prepared to set as our goals in life. Second, we usually do not include satisfaction of our physical needs into the concept, even though we appreciate it and have to achieve it all the time, often simply in order to survive. It would be fair to say that the concept of values describes that part of our goals which are not immediately necessary for survival. We call these aspects of life spiritual, as opposed to the other aspects, which are referred to as physical, or biological.

Goals of organized systems form a hierarchy. When two different goals come into conflict, we call for a higher goal, or a principle, or a value, which we choose to resolve the conflict. The thing which interests us at most today is the highest of these principles: the Supreme Goal, or the Supreme Value of human life. This is the problem of ethics. Philosophy and religion work on this problem traditionally. How does it look from the scientist's point of view? The first attempt of an answer leads to a discouraging result. Science is alien to ethics by its very essence. It answers only to the questions of how things are, but not how they ought to be. It does not say what is good and what is bad. As an American philosopher remarked, no matter how carefully you study the railroad schedule, you will not find there an indication where you want to go.

It is thinkable; however, that science could kill ethics as an independent subject. For somebody who lived in the 19th century and took seriously and consistently the implications of the science of his time, like Karl Marx did, it was quite natural to believe that the problem of ethics was not real, but imagined. In the nineteenth century the picture of the world given by science was broadly as follows. Very small particles of matter move about in virtually empty three-dimensional space.

These particles act on one another with forces which are uniquely determined by their positioning and velocities. The forces of interaction, in their turn, uniquely determine, in accordance with Newton's laws, the subsequent movement of particles. Thus each subsequent state of the world is determined, in a unique way, by its preceding state. Determinism was an intrinsic feature of the scientific worldview of that time. In such a world there was no room for freedom: it was illusory. Humans, themselves merely aggregates of particles, had as much freedom as wound-up watch mechanisms. With this worldview, the problem of ethics is not to decide what is good and what is evil, but simply to predict how people will behave in given circumstances. It is only a branch of science, the science of behavior.

This trend of thinking was the theoretical basis for the Marxist economic determinism, and the Leninist totalitarianism which brought misery and dehumanization to millions, if not billions of people. In the twentieth century the scientific worldview has undergone a radical change. It has turned out that subatomic physics cannot be understood within the framework of the Naive Realism of the nineteenth century scientists. The theory of Relativity and, especially, Quantum Mechanics require that our worldview be based on Critical Philosophy, according to which all our theories and mental pictures of the world are only devices to organize and foresee our experience, and not the images of the world as it "really" is. Thus along with the twentieth-century's specific discoveries in the physics of the micro world, we must regard the inevitability of critical philosophy as a scientific discovery — one of the greatest of the twentieth century.

We now know that the notion that the world is "really" space, in which small particles move along definite trajectories, is illusory: it is contradicted by experimental facts. We also know that determinism, i.e. the notion that in the last analysis all the events in the world must have specific causes, is illusory too. On the contrary, freedom, which was banned from the science of the nineteenth century as an illusion, became a part, if not the essence, of reality. The mechanistic world-view saw the laws of nature as something that uniquely prescribes how events should develop, with indeterminacy resulting only from our lack of knowledge; contemporary science regards the laws of nature as only restrictions imposed on a basically non-deterministic world. There is genuine freedom in the world. When we observe it from the outside, it takes the form of quantum-mechanical unpredictability; when we observe it from within, we call it our free will.

We know that the reason why our behaviour is unpredictable from the outside is that we have ultimate freedom of choice. This freedom is the very essence of our personalities, the treasure of our lives. It has given us as the first element of the world we come into. Logically, the concept of free will is primary, impossible to derive or to explain from anything else. The concept of necessity, including the concept of a natural law, is a derivative: we call necessary, or predetermined, those things which cannot be changed at will. Thus the modern philosophy of science leaves ethics separate from science, and, of course, extremely important, because the kind of life we have depends on the kind of goals we set. Science gives us knowledge, but does not immediately direct our will.

The gap separating knowledge and will can never be fully bridged. It is true and important that knowledge can direct will, make certain decisions natural, highly probable or almost inevitable. But there is no necessity on the path from knowledge to action. With any given knowledge we are still free to set any goal at will. Goals can be logically derived only from goals, not from knowledge. Then is there any way in which science is relevant to ethics? The link between the two is provided by the concept of evolution, and by the inborn feature of human beings which call the will for immortality.

One of the important points to convey is that modern science is a human creation, a very important component of human culture. In that context it is well to appreciate that it is very young - no more than three and a half centuries old - much younger than say drama, sculpture, music, poetry, painting and the like. But already in its brief life science has given us an amazingly rich and wide view of nature, a great deal of it in fact being a result of 20th century science. And this in turn invites us to a reassessment of our place in nature. Another important point is that one must carefully distinguish between science on the one hand, and its exploitation or practical applications on the other. But this is not easy.

The two are of course closely related, one often being the inspiration for the other, yet they must not be mistaken for one another. Erwin Schrodinger, one of the most eloquent writers among the scientists of the 20th century, posed in his book "Science and Humanism" the question "What is the value of scientific research?" and went on to say: "A great many people, particularly those not interested in science, are inclined to answer this question by pointing to the practical consequences of scientific achievements in transforming technology, industry, engineering etc., in fact in changing our whole way of life beyond recognition in the course of less than two centuries, with further and even more rapid changes to be expected in the time to come". But he is not satisfied with this, as he immediately declares: "Few scientists will agree with this utilitarian appraisal of their endeavour".

Then after repeating the question in the form, "What, then, is in your opinion the value of natural science?" he answers: "Its scope, aim and value is the same as that of any other branch of human knowledge it is to obey the command of the Delphic deity, get to know yourself" Science is an unravelling and appreciation of the workings of nature based on two principles which themselves seem not to be derivable from something more deep - that Nature is lawful, and that with patient careful effort we can comprehend that lawfulness. But this is justified by what S. Chandrasekar once said; to the effect that even a mediocre talk or essay can be saved by great quotations. By the way he too used them extensively in his beautiful writings. And when you hear a well chosen piece you will surely feel - how well it is expressed, I wish I had said it myself!

The Beginnings of Modern Science

Let me start with a very rapid recollection of the beginnings and early growth of modern science. The ground work was of course laid by people like Copernicus, Tycho Brahe and Johannes Kelper. But the two main figures who really launched modern science were Galileo Galilei and Isaac Newton, together straddling the 17th century. They were concerned with mechanics or the description and analysis of the motion of material bodies, with the laws of terrestrial and then universal gravitation, and also with optics or the science of light. It was Galileo who declared that the language of nature is mathematics - he said the book of nature is written in mathematical language, and we must learn that language to understand nature. By their joint efforts the principles of modern science based on observation, controlled experiment, analysis using mathematics, concept and theory making, prediction and testing were clarified.

These make up the fundamental methodology of modern science, and they set the pattern for progress in the succeeding centuries. A very important factor in the background which helped them along was a new philosophical attitude which encouraged independent enquiry and observation of phenomena. In this context I always like to recall the beautiful description of Galileo and Newton given by Max Born: "The distinctive quality of these great thinkers was their ability to free themselves from the metaphysical traditions of their time and to express the results of observations and experiments in a new mathematical language regardless of any philosophical preconceptions". During the 18th century there was tremendous progress in applying Galilean-Newtonian principles to wider and wider groups of phenomena.

To a large extent it was also a mathematical elaboration and expansion of what they had initiated. Some of the most illustrious names are the Bernoullis, Euler, Lagrange and Laplace - each of them a mathematician as much as a physicist. Towards the end of the century the sciences of electricity and magnetism also fell into the Galilean-Newtonian pattern and then start the beginnings of modern quantitative chemistry; and soon after the wave theory of light, originally proposed by Huyghens and revived by Thomas Young. Actually the successes of science were already so impressive that they led to a kind of overconfident reaction in more than one way. One was the claim of complete determinism as expressed by Laplace - if we know the positions and velocities of all bodies in the universe at a certain time, and also the forces acting on them, then on the basis of Newton's laws the future would be completely determined. Another was the assertion that in the social sphere too everything was completely lawful - "culture is governed by laws as exact as those of physics.

We need only to understand them.. to keep humanity on its predestined course to a more perfect social order ruled by science and secular philosophy. These laws can be adduced from a study of past history". Both of these claims have long since been understood as being excessive. Science is much more modest today. The third reaction was more philosophical in content; it was in fact an attempt by Immanuel Kant to explain why the physics of Galileo and Newton was so successful. His idea was that some of the fundamental principles of Newtonian physics are unavoidable and inevitable ingredients of the way we understand Nature.

He included the natures of space, time and Euclidean geometry, the law of causality, Newton's Third Law of motion and even universal gravitation, in the list of so-called synthetic a priori categories of thought. They were imposed by the human mind on nature; and rather than being the results of empirical discovery, nature had no choice but to obey them. A really profound understanding of the situation came much later in mid 20th century, and it involved the theory of biological evolution in a very fundamental way.

Science in the 19th and 20th Centuries

The claim that social systems obey strict laws of the kind present in science led to a predictable romantic reaction. One consequence was that then science was pretty much left to itself, which helped it to make continued progress. I can only highlight the most important achievements, first in the 19th and then in the 20th century. The sciences of electricity and magnetism matured, the concept of the electromagnetic field was created, and finally in Maxwell's hands in 1865 the understanding of light became part of electromagnetism. Then thermodynamics followed by statistical mechanics grew in the work of Carnot, Clausius, Kelvin, Maxwell and Boltzmann. In biology the theory of evolution by natural selection was put forward by Charles Darwin in 1859, while in chemistry the great systematization of the elements was achieved by Mendeleev a bit later. In the closing years of the 19th century came a string of discoveries - spectroscopy, x-rays, the electron, radio activity - that would profoundly influence 20th century physics. Let me interject something about technology at this point, especially in the 20th century.

As Schrodinger said, the changes due to technology have profoundly altered the patterns of life in most parts of the world, though it is of course unevenly so. In health, food production, communication, travel and entertainment - to mention only the most obvious areas - there is no comparison between conditions in the early 1900's and now. Much of the technology of the first half of the 20th century was based on 19th century Maxwellian physics; only towards the end of the 20th century have we seen the technology resulting from earlier 20th century science. I would like to quote from Freeman Dyson at this point: "It usually takes fifty to a hundred years for fundamental scientific discoveries to become embodied in technological applications on a large enough scale to have a serious impact on human life.

One often hears it said that technological revolutions today occur more rapidly than they did in the past... In reality, the time elapsed between Maxwell's equations and the large-scale electrification of cities was no longer than the time between Thomson's discovery of the electron and the worldwide spread of television or between Pasteur's discovery of microbes and the general availability of antibiotics. In spite of the hustle and bustle of modern life, it still takes two or three generations to convert a new scientific idea into a major social revolution". Within 20th century physics we have seen vast conceptual changes through the two relativity theories and then through quantum theory.

The understanding of the nature and relationships between space, time, motion and gravitation has been profoundly altered. And from quantum mechanics we have learnt that the physics of microscopic objects - at the molecular, atomic, nuclear and sub nuclear levels - is utterly unlike what had been learnt earlier through phenomena at our own scale. Quantum mechanics has also supplied the real basis for chemistry, the understanding of the periodic table, chemical structures and reaction mechanisms. As for the physics of stars, galaxies and the universe the changes in understanding have been unbelievable in magnitude. Even as late as 1919 it was thought that the universe consisted of just our Milky Way galaxy. Observations and theory going hand in hand have now disclosed that our galaxy is just one of 1011 galaxies, each containing 1011 stars; and there are fascinating structures at all scales.

Turning to biology, partly helped by the tools supplied by physics and chemistry, the basic molecular structures and processes underlying all of life have been unravelled; and the unity underlying the amazing variety of living organisms, as well as the molecular basis for Darwinian evolution to operate upon, have become evident. No wonder then that the 20th century is being called the Century of Science.

The Character of Science

I emphasized at the start that science is a human creation. There are several aspects to this, all worth description. Science is an unending and continuous exploration of nature, it is always unfinished. As we learn more and more, new questions continually arise; and we exploit knowledge already in our hands to probe further into finer and finer details of phenomena. Sometimes the important questions then change, old ones are transcended and lose meaning. And even answers when given are provisional and may undergo revision as we progress and in Dyson's words: "Science is not a monolithic body of doctrine. Science is a culture, constantly growing and changing". Physics offers splendid examples that illustrate this. Before Newton found his Law of Universal Gravitation, it was generally believed that material bodies could influence one another only by direct contact. But his law was a law of action at a distance. He himself was uneasy about this and said so in a letter to a friend. But his law worked beautifully, and over the next century or so the idea of action at a distance was accepted and woven into the fabric of physics.

Even electrostatics and magnetostatics were initially expressed in this framework. Only later around 1830 came the concept of the electromagnetic field, created by Faraday and perfected by Maxwell; and then action by contact was restored in the description of electromagnetic phenomena. In the case of gravity this was accomplished by the general theory of relativity in 1915. The story of light is another impressive instance of this phenomenon. Huyghen's very early wave conceptions were eclipsed by Newton's corpuscular ideas.

The wave theory had to wait till Thomas Young's experiments around 1800 for its revival. But then came the concept of the ether-waves had to be waves of something howsoever refined! It was a long struggle before the ether was finally given up, thanks to special relativity; only then were electric and magnetic fields and their waves accepted as irreducible constituents of nature not made up of anything simpler. But around the same time came the photon idea-that light is not simply a classical wave as envisaged by Maxwell, but has a particulate aspect or graininess as well. It is the reconciliation of the wave and particle natures that is achieved in quantum theory. Science grows in small steps, with built in self correcting mechanisms.

It is made up of tested and corroborated and hard-won knowledge, which is given up or modified only when new evidence compellingly demands it - a fine combination of conservatism and the willingness to change. The fact that it is a collective enterprise was sensitively captured by Rutherford: "Science goes step by step and every man depends on the work of his predecessors. When you hear of a sudden unexpected discovery, a bolt from the blue, as it were you can always be sure that it has grown up by the influence of one man on another, and it is the mutual influence which makes the enormous possibility of scientific advance. Scientists are not dependent on the ideas of a single man, but on the combined wisdom of thousands of men, all thinking of the same problem and each doing his little bit to add to the great structure of knowledge which is gradually being erected". It is the dependability of this carefully tested cumulative knowledge that is the strength of science that gives us the confidence that Nature is comprehensible and that new properly posed questions will ultimately be answered.

Science and Art

When we compare the two components of culture - science and art, we see both similarities and differences. In both, the importance of training, tradition and discipline before one can make creative contributions cannot be overemphasized. Even when there is a break from tradition, that tradition is crucial; it supplies a point of departure, something to break away from! And as for discipline, as science grows in extent the training needed before one can contribute keeps growing. Of course there are differences: science seeks to understand and explain outer objective reality, while art is concerned with inner emotional subjective responses to human experience as well as to the external world. I mentioned the cumulative nature of scientific knowledge. In a special sense the role of the individual gets altered as compared to art. Individual discoveries become parts of the whole; they are often reformulated for easier understanding, and in time are subsumed in later developments.

One way of referring to this process is to speak of "the inevitability of scientific discovery". In a sense ultimately it does not matter who discovered a particular effect or law, though in remembering the historical development of science we do associate names with discoveries. It is often said that if Einstein had not formulated special relativity in 1905, someone else would have done so sometime soon. Though, to be honest, in the cases of the general theory of relativity and the Dirac equation the stamp of Einstein and of Dirac seems very difficult to erase! Similarly Crick has said that if he and Watson had not discovered the structure of DNA in 1953, someone else would have done so within two or three years. Contrasted against this is the uniqueness of every artistic creation, even when it is embedded within a tradition. Each artistic work is the expression of an individual creative spirit; it has wholeness to it and it makes no sense to reformulate or re express or explains it for easier comprehension.

The themes that underlie artistic creativity are very ancient, much older than modern science; they are rooted in human emotional and social experience and in responses to nature. Of course the medium of expression does change now and then. And most often in the finest works of art the communication between creator and audience is immediate and transcends logic and analysis. But when we look at the creative processes in science and in art the similarities come back. After all each brick in the structure of science is the result of the effort of some individual (or these days more and more often some group); that is just the sense in which science is a human creation, except that it has to stand the test of comparison against impersonal nature and be repeatable.

Now the point is that scientific discovery too often results from inspired imagination going beyond logic and reason, bold leaps of thought; though for later communication and verification it has to be expressed in sequential logical steps. From all accounts there is a thrill experienced at the moment of scientific discovery which is very close to the emotions that accompany artistic creation. In physics in particular, during the 20th century, the roles of aesthetic criteria and of sensitivity to deep beauty have come through very sharply. These are of course subjective statements but they do influence the outlook of working physicists, even of those who can only admire the work of others. To sum up this comparison and help close a seeming gap let me turn to Einstein: "The most beautiful experience we can have is the mysterious. It is the fundamental emotion which stands at the cradle of true art and true science".

The Problem of Values

Not only has modern science revealed to us the immensity of the universe - both at the macroscopic and at the microscopic levels - it has also shown us how small a part of the whole display we are. As far as we can tell, life has occurred just once on earth, on one small planet circling the sun, which is in turn just one of 1022 such stars among all the 1011 galaxies making up the universe. Even this singular occurrence of life has been the result of a chance event. And the long chain of evolutionary steps leading from the earliest forms of life to us today seems to be the result of blind chance at every stage, acted upon by natural selection. We are in no sense the centre or purpose of it all. It turns out that we are subjected to three learning processes, involving very different time scales.

The slowest is phylogenetic learning - this takes place over millions of years, as species evolve governed by natural selection. It is here that the abilities to perceive the most important features of the world around us - at our own scales of size and time - and to survive in it, are selected for and perfected. Naturally this gives us the direct ability to sense only a very tiny fraction of what surrounds us; and our commonsense notions and intuitions and primitive scientific concepts spring from this domain. The limitations of this domain were expressed by Emil Wiechert very perceptively long ago in 1896: "The universe is infinite in all directions, not only above us in the large but also below us in the small. If we start from our human scale of existence and explore the content of the universe further and further, we finally arrive, both in the large and in the small, at misty distances where first our senses and then even our concepts fail us".

We see here the basic explanation of the origin of the synthetic apriori of Kant: they are the results of slow phylogenetic learning, and are understandably limited in scope. We also see why in the physics of the very large and of the very small we have to depend so enormously on mathematics to guide us. The second learning process occurs through the course of human history, over a few thousand years. Here the dominant determining factors are cultural. Religions and religious movements really belong to the 'childhood phases' of human history, and through science we learn that we must go beyond. It is interesting to learn that the religious instinct has genetic roots, while the ethical sense is much older than religion and comes from our evolutionary heritage.

Nature is not to be feared but to be understood rationally. Einstein often argued for a new meaning and interpretation for religion - the recognition that there is a central order in Nature, which is accessible to our understanding. No need for an anthropomorphic God responsive to our individual pleas and prayers. The third learning process - ontogenetic learning - takes place within each individual life time. Here we are born with the abilities slowly fashioned and given to us by phylogenetic evolution, but we think they are innate and inevitable. As we grow we put to use this limited perceptual apparatus and learn about the world around us; and in science we go far beyond our senses to learn about phenomena at far larger and far smaller scales than ourselves. All the while we are on a 'genetic leash' - our behavioural and thinking patterns are strongly influenced by our evolutionary past in ways we may be reluctant to admit!

In this entire where do we turn to for values? The first important lesson is that in science, in knowledge about the working of nature, there is no value, no ethical sense, in human terms. We have to search for values within ourselves. Quoting from Bertrand Russell: "Science, by itself, cannot supply us with an ethic. It can show us how to achieve a given end, and it may show us that some ends cannot be achieved". We have to learn to use wisely the knowledge that we generate. As far as technology is concerned, Dyson said it concisely: "Technology without morality is barbarous; morality without technology is impotent" A more extended exhortation comes from Robert Pirsig who says in his book "Zen and the Art of Motorcycle Maintenance" of a quarter century ago: "The way to solve the conflict between human values and technological needs is not to run away from technology. That's impossible. The way to resolve the conflict is to break down the barriers of dualistic thought that prevent a real understanding of what technology is - not an exploitation of nature, but a fusion of nature and the human spirit into a new kind of creation that transcends both".

And later: "The place to improve the world is first in one's own heart and head and hands, and then works outward from there". Dyson's aphorism seems patterned after the familiar statement by Einstein which goes back to the relation between science and religion: "Science without religion is lame; religion without science is blind". A great deal of truth captured in a few words. The reference here is to the collective aspects of the situation more than to the individual and also a pointer to the great responsibility that the scientist bears - to communicate to others the understanding of the lawfulness of nature and the open questions, and to create the climate where we can assess ourselves and understand the role of the thinking and informed individual. But to my mind the deeply human aspects of the situation at the level of the individual are best expressed in my third and final aphorism taken from Victor Weisskopf and accompanying the previous two from Dyson and Einstein: "Human existence is based upon two pillars: compassion and knowledge. Compassion without knowledge is ineffective; knowledge without compassion is inhuman".

2

Alternative Views of Technology

Alternative technology is a term used by environmental advocates to refer to technologies which are more environmentally friendly than the functionally equivalent technologies dominant in current practice. It is technology that, as an alternative to resource-intensive and wasteful industry, aims to utilize resources sparingly, with minimum damage to the environment, at affordable cost and with a possible degree of control over the processes. The term is sometimes confused with appropriate technology, but while there is significant overlap, the terms have different meanings, particularly related to the importance of low cost and ease of maintenance for developing country applications.

Alternative technologies themselves are part of environmentalist politics. Common political issues related to alternative technologies include whether they are practical for widespread use; whether they are cost-effective; whether widespread adoption would produce negative impacts on the economy, lifestyle or environment (production energy costs/pollutants); how to encourage rapid adoption; whether public subsidies for adoption are appropriate; which technologies government regulations should favor, if any, and how environmentally unsound technologies and practices should be regulated; what technological research should be done and how it should be funded; and which of a field of competing alternative technologies should be pursued.

The term was coined by Peter Harper, one of the founders of the Centre for Alternative Technology, North Wales (aka *The Quarry*), in Undercurrents (magazine) in the 1970s. Some "alternative technologies" have in the past or may in the future become widely adopted, after which they might no longer be considered "alternative" and for example the use of wind turbines to produce electricity. Technology is generally accepted to mean "the practical application of science to commerce or industry". The transfer of technology from rich to poor countries is often argued as a necessary condition for the poor to industrialise. Often this is also used to persuade Developing Countries to accept liberal rules to the flow of foreign capital – mainly in the form of foreign direct investments (FDIs). The reality is that very little technology transfer actually occurs, or if it does, it is in the form of Psuedo Transfer.

There is, however, an alternative view on technology that puts the matter in a more realistic perspective. It says that an alternative technology approach must be based on three policy strategies, listed below in order of priority:

a. Creation of a home-based Domestic Scientific and Technology Capacity (DSTC), including capacity to undertake relevant research and development.
b. Purchase (as opposed to transfer) of appropriate technology from the open market.
c. Transfer of technology, preferably between South-South, only under certain conditions.

Technology and Technology Transfer

Mainstream economic literature on technology falls into two main theories. One approximates a kind of technological determinism, which argues that with the "right" kind of policies, the developing countries can "attract" the right kind of technology, which would then surge them onwards on the road to development. The second looks primarily at the social, political and institutional aspects of technology. Technology determinism is when technology is given a determinist role in production and in economic development. Very few development economists are out and out technology determinists, but many approximate the description when they give it a decisive role. A good example of this is the development economist, Sanjay Lall. He argues that integration into the process of globalisation offers the best prospect for the developing countries.

"As embodiment of technological progress and more open markets," argues Lall, "Globalisation offers enormous potential productive benefits to the developing countries." The gap between rich and poor countries will diminish if the latter take "correct policy decisions". For example, a country that wants to expand exports must upgrade its technology structure (the "market positioning" argument). They have to move into higher level manufacturing, design, development and regional service activities. In terms of market shares, primary products have been losing ground steadily since 1976. They must therefore get out of primary production, and aspire to get into medium or high technology based production for exports. He gives the examples of ten developing countries that have done well through this route – namely, China, Korea, Taiwan, Mexico, Singapore, Malaysia, Thailand, Brazil, the Philippines, and Indonesia.

Argentina has also been cited in the past as a success story of mainstream strategy, but with its demise in recent years it has dropped from the list. Six of the ten above listed countries - Korea, Mexico, Thailand, Brazil, the Philippines and Indonesia - are now experiencing serious reversals in terms of economic and social development, and should also be dropped from the list. In spite of this, there is such a seductive appeal to the mainstream argument that many developing countries buy into it. The New Economic Partnership for Africa's Development (NEPAD) is one example, primarily because its authorship draws primarily from this genre of development economists. NEPAD lays an extraordinary emphasis on technology (especially Information Technology) as a means of Africa leapfrogging into a new era of development.

In order for this to happen it prescribes the creation of the "right conditions" for attracting foreign direct investments (FDIs), which it argues will bring technology into Africa. This, it maintains, will create for Africa a niche in the global "value chain" in which it will have a competitive edge over other players. A more discriminate theory within the mainstream looks at the social, political and institutional aspects of technology. A good example of this is Carlota Perez. She looks at the entire "meta-paradigm" of technoeconomic change affecting the economy - including production systems, organisation, property regimes, social, political and managerial change, and interdependence between all these factors. In this gamut of relationships, Perez singles out the link between technology and finance capital.

She argues that the full fruits of technological revolution occur roughly every 50 years, widely reaped with a time lag of 20-30 years of turbulent adaptation and assimilation. She explains why, in spite of the uniqueness of each period, there is a certain sequence of events that occurs every half century. This is at the global level. At the level of the developing countries, Perez identifies finance and debt as the decisive factors for technological innovation or technology "diffusion" (not transfer). Countries and regions vary in their capacity and desire to make required changes, depending on sociopolitical factors and historical circumstances. Old institutions have to be replaced by new ones because of mismatch (decoupling) between techno-economic change and socioinstitutional development.

Alternative View on Technology

An alternative view comes closer to Perez rather than NEPAD-Lall analytical perspective. FDIs may or may not bring technology to their new areas of conquest. This depends on many factors - such as its cost, the availability of local technology and skills, the financial and institutional factors, and the degree of control the owners of technological knowledge retain over the technology. The degree, to which an actual "transfer" takes place, also depends on many factors – crucially the host country's domestic scientific and technological capacity (DSTC), and the terms of the negotiations between the owners of the technology and its "recipients". The role of domestic policy in facilitating industrial clusters which benefit from each other's technology, markets and skills, is well known by industrial thinkers and practitioners. In other words, technology does not just come like that – with the drop of an FDI hat (or begging bowl). Technology, above all, is a means of control over both production and market.

Technology is not neutral. It is embedded in capital, and it is controlled by whoever controls the capital. Capital and its embedded technology enable its owners to make policy on production, marketing, pricing and the distribution of income. It is an illusion to think that developing countries will secure this kind of embedded technology from the West and then out-compete the West. Countries such as Korea and Taiwan, as all other now advanced economies in history, were able to do it because they disembodied the technology from its capital base (by, for example, copying intellectual property, and through reverse engineering), and by creating a "national" base for capital. Some countries were able to do this during the cold war years when the West needed them to fight against the Communist threat coming from China and Vietnam.

In countries like Korea and Taiwan the local entrepreneurs created their own national companies (national "champions"), generously supported by their governments (through credit facilities and trade barriers against foreign goods), that first secured the domestic market and then entered into competition with the Western companies in the global market. Since the end of the cold war, this option is no longer available. Now that the cold war is over, even Korea cannot expect indulgence from the United States. The US, Europe and Japan took the first opportunity that availed itself with the financial crisis in Korea in 1997/98 to assert their control over the industrial and financial empire that Korea had so painstakingly created over thirty years of hard work of its workers and the skill of its entrepreneurs.

Now, with intellectual property rights embedded in the World Trade Organization (WTO) under the Trade-Related Intellectual Property Rights (TRIPS), scientific knowledge has become monopolised in the hands of a few thousand multinational corporations that use this knowledge to control the economies of the third world. High technology has come in areas such as mining (especially oil, gas and strategic minerals) and services (banking, finance, etc.), only because these have given the foreign direct investors control over these sectors. All the hype about creating the "enabling environment" for FDIs on the grounds that they would "bring" technology must be seriously discounted. China is often cited (as by Lall) as an example, but the Chinese case does not bear out the liberalisation paradigm.

The Chinese know that the western corporations have to export capital (and therefore also technology) for their own reasons. They know that foreign capitalists will try to use technology to control production and market in China. So they negotiate hard, and play one Western or Japanese company against another, until they get what they want. They do not accept FDIs – which is a packet of money capital, technology, management, market access, and much else. They disaggregate the package, and take only what they need (for example, the Chinese eschew money capital; they have plenty of their own), and try to maintain control over domestic production and market. They also insist that the foreign investor transfer not just application end-use technology (for example, mobile phones) but also design technology and the knowledge that comes with it.

In this light, the alternative perspective makes a distinction between three kinds of technology transfer. Adaptive Transfer takes place where Foreign Technology (FT) supplied in year one is adapted by a domestic technological and scientific capacity (DTSC) before going into Production (P) in year two or three. Full Transfer takes place where FT is purchased in year one, is simultaneously used in production and made the subject of domestic Research and Development (R&D) and engineering design, and in the year 'N', the technology is renovated or upgraded so that local/national DTSC is able to deliver the renovated or so-called 'next generation' technology. And Pseudo Transfer takes place where FT functions only as an input into production, with no impact on DTSC.

The Chinese insist, as far as possible, on full transfer of technology, even if it goes through stages of adaptive transfers. In Africa, by contrast, most of so-called "technology transfer" is of the third kind. Pseudo-transfers of technology are essentially excuses for transnational corporations (TNCs) to take over local companies, or to carve out a share of the domestic markets. The above typology is offered as an alternative to the taxonomy in mainstream development literature. Instead of breaking down technology in terms of the absorptive, adaptive or transformative capacity of the host country, it does so in terms the sophistication of the technology. Thus, for example, Sanjay Lall breaks it down into low, medium and high technology. He argues that the developing countries should shift from low level to either medium level or high level technology.

The Chinese success, however, is not based on what the above taxonomy of mainstream development economics might suggest. Indeed, the Chinese are buying "low level" discarded machinery from Western countries at virtually the cost of scrap for the older industries, such as ginning, textiles, and a whole array of consumer goods. (A Capstain lathe machine, for example, that cost $100,000 some 15 years ago can now be purchased for as low a price as $100 in the auction floors in Europe). In the older industrialised countries, the cost of labour is so high that it is no longer economical to employ these machines, even though the quality it provides is fully comparable to the computerized products. They have moved on to semi-automatic, then automatic and now computerised technology in order to beat the wage cost. They have moved up the "technology ladder" in order to retain their competitiveness.

The Chinese, on the other hand, can purchase the discarded machinery at very little cost, reassemble them at home, employ labour at the prevailing rate in the national market (disparagingly called "cheap labour" by the West), and then export quality products to the Western markets at prices that are one-tenth or one-twentieth of the production cost in the West and at the same time, the Chinese are developing their own DSTC. They send their young graduates to the US and other advanced countries in order to learn science and its application to production, and then ensure that they (or most of them) return to China. It is in this specific historical and institutional context that China crafts its policy on research and development.

Options for Africa

For Africa the Chinese experience provides a good lesson. However, Africa should not try to compete against China in the consumer goods export market. The wages in Africa are much higher than those in China. It is better for Africa to buy relevant technology off the shelf (or in auction markets), than to ask for technology "transfers", especially since the transfer aspect is seized as an opportunity by Western enterprises to enter and control African production and markets. If any "transfers" are needed (and these cannot be excluded), then it is better for Africa to negotiate these with other developing industrialised countries than with the highly industrialised countries of the West. Africa should not use the latest technology for mass production. Africa needs labour-intensive, slow moving machines that can produce enough goods to satisfy the domestic and regional market.

Even if Africa is able to mass produce using the latest say laser beam technology and computerised machinery, where will they sell the products? The domestic market is limited, and in the export market they cannot compete either in the labour-intensive products (in which China for example has a competitive edge) or in capital-intensive products. Besides, in order to produce using the latest technology, they will have to import all the necessary technical and managerial skills from outside, which make the whole enterprise non-viable. Indeed worse, because in the process it creates debt finance and an increasing burden on debt servicing. It is in this context that Africa must develop its own DSTC, including a policy on relevant research and development.

The R&D policy must be based on the production condition in the region, the need first to produce for the domestic/regional market (only secondarily for the export market), and Africa's location within the global value chain. This can be done even if the country is as small as Cuba and has little by way of natural resources. Cuba has its own research and development strategies in the areas, for example of pharmaceuticals, and the sugar and tobacco industries. Investment in scientific education and engendering a culture of innovation based on indigenous knowledge, alongside policies which demand such development and domestic control over it is clearly the principal route to developing science and technology.

The above strategy is also recognised in the UNDP publication Making Global Trade Work for People. It says

"...the experience of industrial countries and successful developing countries provide two other important lessons. First, economic integration with the global economy is a result of successful growth and development not a prerequisite for it. Second, domestic institutional innovations many of them unorthodox and requiring considerable policy space and flexibility have been integral to most successful development strategies."

To their national efforts, it may be added in conclusion, the DCs and LDCs of the countries of the South can add regional efforts to pool resources to try and build regional DSTCs. In the SADC region for example, a country like Zambia or Mozambique can take advantage of the superior DSTC of South Africa, or the skills of Mauritius in niche sectors (for example, in the sugar industry), at the same time as they take advantage of their status as LDCs in international trade agreements. SADC and/or COMESA should develop a protocol on the subject that identifies modes of technology transfers between member countries. They should also strategise on the building of a regional DSTC without having to depend on foreign TNCs or FDIs.

Conflicting Visions of Technology

The mentions are inevitably going to be far too brief; everyone of them deserves intensive thought and discussion and they're not at all comprehensive. But these are some of the threads that ought to be woven together to give some kind of coherent picture of where we stand today, what kinds of problems we have to face and where we might find at least a standpoint to begin to think about them in a constructive way. There are occasional moments in human affairs where power relations make it possible to establish social and economic arrangements that actually merit the term world order. Merit might not be the right word. It's not necessarily a phrase that should be invested with positive connotations, as history amply reveals. One of the most dramatic and in fact most easily timed of those moments was about 50 years ago in the aftermath of the most devastating single catastrophe in human history which took place right in the heartland of western civilization.

At the end of the war, the United States, of course, had an overwhelming share of global wealth and power and, perfectly naturally, dominant forces within the state corporate nexus in the United States planned to use that power to organize the world as much as they could in accord with their own conceptions of their interests and those they represented. Of course, there were conflicting visions both at home and abroad and they had to be contained, or better, rolled back, to borrow some cold war rhetoric. That was done with varying degrees of success, but in fact the basic conflicts persist and for elementary reasons - they persist because they are about fundamental values; they are about freedom and justice and human dignity and human rights in a world of inequality - great inequality and great concentration of power (the real world, that is); these values quite commonly constitute an arena of conflict between centres of power and most of the rest. A good deal of history revolves around these conflicts in the last half century; this is no exception and the next will not be either.

Well at the onset of the current era, about a half century ago, the framers of the world order of that day, the new world order of that day, they faced these challenges everywhere. At home, what had to be contained, maybe rolled back, were the very strong commitments of a large majority of the population to social democratic ideals that the business world rightly perceived as a grave threat to their traditional dominance. They were the hazard-facing industrialists in the rising political power of the masses, as the National Association of Manufacturers put it in their internal literature. It was the crisis of democracy that was posed by a population that sought to enter the political arena, as frightened liberal internaltionalist elites phrased essentially the same problem after the ferment of the 1960s, expressing particular concern about what they called the institutions responsible for indoctrination of the young, which were failing to carry out their disciplining role properly.

Similar problems were faced throughout the industrial world. They were enhanced by the prestige and appeal of the anti-fascist resistance, which was a complex affair, but often had radical democratic thrusts. They were enhanced further by the discrediting of the traditional conservative order, which had been linked closely to the fascist system. Reinstating that traditional order in its essentials was a primary task of the early post-war years and it was achieved to a large extent often in not very pretty ways. As in the United States, this project continues. It's taken new forms in the last 25 years, as here and as throughout the world, under the guise of neo-liberalism or economic rationalism or free market doctrine, which is permeated with a good deal of deceit, hypocrisy and maybe outright fraud. All of these issues are strongly very, very much alive right now - here, Europe and elsewhere.

In the Third World, the south - the developing world as its euphemistically called - similar problems were compounded by strong pressures, uncontrollable pressures, to overturn the imperial systems, and the legacy of dependency and subordination that they had left. The basic issues were very much the same in most of the world but they were revealed with particular clarity, with starkest clarity, in Latin America for the simple reason that the United States faced no challenge there, no outside challenge, so you see the principles operating in their purest form. There was a real challenge, but it was from the domestic population - no outside challenge.

As In Europe, these conflicts in Latin America came to a head even before the war was over, in the case of Latin America very dramatically in February 1945 at a hemispheric conference. The hemispheric conference was supposed to organize affairs for the hemisphere. We know from U.S. internal records now that the United States was deeply concerned with what the State Department called the "philosophy of the new nationalism" that was spreading all over Latin America and indeed all over the world, quoting State Department documents. The philosophy of the new nationalism, which embraced policies designed to bring about a broader distribution of wealth and to raise the standard of living of the masses on the principle that the first beneficiaries of the country's resources are the people of that country. That heresy is called radical nationalism or economic nationalism and of course it has to be stamped out - the first beneficiaries of a country's resources are U.S. investors, their counterparts elsewhere, and local elites who are associated with them.

At the hemispheric conference the United States - given the power relations of course the U.S. prevailed - and it imposed what was called an economic charter for the Americas which called for an end to economic nationalism in all its forms. In the cruel and bloody history of the half-century that followed, these remained central themes and they will continue to. They are very much alive today; now they are often framed differently in the context of the investor rights agreements that are mislabeled free trade agreements: NAFTA, the forthcoming, maybe, Multilateral Agreement on Investments, MAI, and what's called globalization, which is a specific form of international integration (not by any means the only necessary form); it's the specific form that's crafted primarily to serve the interests of its designers, again not terribly surprisingly, transnational corporations, financial institutions and the bureaucracies that they control and of course the major states that are part of the system.

The most critical part of the Third World was then and remains today the Middle East, for the very simple reason that it's the locus of the world's major energy supplies for as far ahead as anybody can see. Hence, it was considered to be, and is still considered to be, of particular importance that the first beneficiaries of that wealth are not the people of the region; rather the resources must be under effective U.S. control, they must be accessible to the industrial world on terms that the United States leadership can see is appropriate and, crucially, the huge profits that are generated must flow primarily to the United States, secondarily to its British junior partner, to borrow the term used by the British Foreign Office rather ruefully to describe its new role in the post Second World War era.

This is done in various ways. In part it's recycled by local managers who have to be dependent on the global rulers, a long story which continues. Well quite naturally these arrangements breed continual conflict. Internal U.S. documents describe them in the conventional way. The conflicts are conflicts with radical nationalism, radical Arab nationalism that threatens U.S. dominance. For the public it's put a little differently, varying over time. These days it's international terrorism, or the clash of civilization; tomorrow it will be something new, but it's basically the same ones all the time. The question is, who's going to be the first beneficiaries of the region's resources.

These conflicts are likely to become more virulent and ominous in the coming years, at least if the analysis and projection of quite a number of geologists are anywhere near accurate. A reasonably broad consensus (there's plenty of room for disagreement and uncertainty) but a reasonably broad consensus was captured in the headline of a major review article on the topic, in the journal Science - the journal of the American Association of the Advancement of Science a couple of weeks ago. The headline was, "The next oil crisis looms large and perhaps close." It may be a little hard to believe in a period when gasoline prices are at an historic low, but there are many who regard that as an aberration, a short-term aberration.

The crisis that many people fear is that the rate of discovery has been declining for some time after having risen steadily since the earliest discovery of oil, and the Gulf region, the Arabian peninsula and the Persian Gulf region, that has by now virtually regained the share of energy production that it had in the early 1970s, (You'll recall that that was sufficient to bring the era of super-cheap energy to a sudden end; it happened to be a temporary end but a foretaste of what lies ahead.) basically back to that share, meaning that degree of power, and that share is expected to increase, in part because world consumption is increasing very rapidly and most of the known energy reserves by a big measure are in that region. And it's also speculated, not apparently implausibly, that something like the 50 per cent mark of exploitable capacity may not be too far away, maybe within the next few decades. All of this combines suggest to policy makers and others that the need to control that region is going to become increasingly important and that's going to mean very likely increasing confrontations with radical nationalism.

As a kind of a sidelight to this, very likely, the latest terrorist exchange in the last few weeks might well be seen in this context. Referring to the terrorist bombings of the U.S. embassies in Africa, allegedly by groups opposed to U.S. domination of the major oil producers, and the U.S. missile attacks on Sudan and Afghanistan. One might ask, why those targets? Well, like the bombings of the embassies in Africa, the U.S. selected targets that were vulnerable, not the ones to which the messages were aimed, in either case. The message for the missile attacks may well have been directed elsewhere, in this case very likely to Riyadh and Teheran. There have been recent steps towards rapprochement between Saudi Arabia and Iran, historic enemies, and that's not an appealing prospect for U.S. global managers.

It raises fears, which have been lingering for a long time, of regional groupings that will get out of control in the strategically most important part of the world, which holds the greatest material prize in world history - that's quoting U.S. assessments from the late '40s, which still prevail. The U.S. missile attacks have been criticized (you've read plenty of criticisms of them) as being counterproductive (elite opinion has held that) because of their effects on the Sudan and Afghanistan. Well, it's a pragmatic judgment, apparently. The same opinion seems to be largely unconcerned by the fact that, effective or not, they were war crimes - that's now partially conceded in the case of Sudan. However, just keeping to the pragmatic judgment, it might be evaluated in the light of a secret 1995 study of the U.S. Strategic Command, called Essentials of Post-Cold War Deterrence, which was released recently under the Freedom of Information Act. It's an interesting document.

It resurrects Nixon's madman theory, as it was called. It says that the United States should portray itself as irrational and vindictive with leadership elements out of control and it should exploit the nuclear arsenal for that purpose. This madman posture can be beneficial to creating and reinforcing fears and doubts among adversaries, real or potential. In this case perhaps the big players in the region, Saudi Arabia and Iran, whose potential rapprochement, which has been going on now for almost a year, is doubtless a very frightening prospect in Washington. Well, we don't have documentary evidence, so that's speculation. But it's not unreasonable. Well there's a lot to say about all these topics and related aspects of the post-war global system that haven't even mentioned. But just leave it as something to think about and turn to something related: namely, the institutional structures, the institutional framework that was designed for world order 50 years ago, and how it's fared, and what it looks like today.

The institutional structure had three basic components. One was an international political order - that's articulated in the United Nations Charter. A second part was concerned with human rights and the norms for the behavior of governments towards their citizens - that's the Universal Declaration of Human Rights. The third component was an international economic order. That's the Bretton Woods system, as it's called, designed by the United States and Great

Britain, the global manager and the junior partner. The first international political order (essentially the UN Charter) based on a very simple, straightforward principle and elaborated in various ways. The principle is that the threat or use of force in international affairs is disallowed, unacceptable, has to be barred totally, with two exceptions.

One exception is when the threat or use of force is specifically authorized by the Security Council after the Security Council determines that peaceful means have failed. The second is the famous Article 51, which says that nothing in this Charter abrogates the right of self-defence against armed attack until the Security Council acts. That's a rather narrow and specific notion. It means, for example, that if Cuban armies invade the United States, the United States is supposed to notify the Security Council and then, until the Security Council has a chance to do something about this terrible threat, it's allowed to defend itself in anyway that's necessary. Whether that example is hypothetical or not depends on what you want to believe. The Cuban threat to the United States was recently downgraded by the Pentagon, so we don't have to tremble in total fear anymore.

That elicited a good deal of anger in Congress and the conclusion was rejected by the White House, which invoked Cuba's threat to the national security and existence of the United States, just a few months ago, in the course of rejecting World Trade Organization jurisdiction when the European Union protested before the WTO gross U.S. violations of trade agreements and international law that had already been condemned by just about every international agency, including even the normally quite compliant Organization of American States. So, depending on where you sit, that example was real or hypothetical, but whatever it is, that's the exception permitted by the UN Charter by the framework of political order and received international law. And those are the only exceptions to the threat or use of force.

Now of course there is no enforcement mechanism -

this has to be by acceptance. There is in fact an enforcement mechanism, namely the great powers, and to be realistic exactly one of them, namely the United States, so that's the enforcement mechanism. But that suffices to show that the whole system is null and void because the United States rejects the principles out of hand. It rejects them both in practice and in fact in doctrine. There's no need to waste time on the practice in the past half century; the bombing of a pharmaceutical plant in the Sudan a couple of weeks ago is a recent illustration but one that's completely trivial in historical context, though terrorist destruction of half of the medical supplies and fertilizers in the United States might be taken a shade more seriously.

Whatever the practice may be, and you can know that perfectly well, or should, what's interesting about the recent years is that official doctrine expresses with great clarity and precision the utter contempt for the principles of world order that are of course grandly proclaimed when they serve some power interest. This has been going on since the Reagan years and it is a change, a doctrinal change; you might say it's a change toward greater honesty. Anyhow, it's a change. Since the Reagan years the United States has officially reinterpreted the Article 51, the crucial article, to justify its repeated reliance on force. It has held that Article 51 authorizes self-defence against future attack. Article 51 permits the United States to defend its interests. This became even more ludicrous in the Clinton years. It was all formulated rather straightforwardly by Ambassador Albright, now Secretary of State, when she informed the UN Security Council, which was then refusing to go along with some U.S. demands about Iraq...she informed the Security Council that the United States will act "multilaterally when we can, unilaterally when we must in an area important for our interests."

That means unconstrained by the world court which had already been dismissed as irrelevant 10 years earlier by the most solemn treaty obligations, the foundations of world order, and so on. The only innovation in all of this in the past 15 years is that contempt for these high principles is now openly proclaimed, with the acquiescence and the applause of the educated classes - that's a change. But it serves to indicate where the foundations of world order stand after 50 years. In brief, the United Nations and its Charter are fine when they serve as an instrument of power, otherwise the decisions and the condemnations are not even worth reporting, and they are not reported, let alone obeying.

Incidentally, you've been following the recent debate about founding an international criminal court on war crimes and as you know the United States, essentially alone refused to go along with that. U.S. opposition is effectively a veto when the General Assembly votes 151 to 1 on something (or 2 if the U.S. picks up a client state). That amounts to a veto, just for straight power reasons - nothing obscure about it. And this effectively vetoes the criminal court on war crimes. The official argument that was given by the Clinton administration and Congress was that an international tribunal might carry out frivolous prosecution of U.S. soldiers engaged in peacekeeping operations. That's not very credible, especially if you look at the U.S. role; the U.S. is mostly disqualified from peacekeeping operations - that's literally true.

The reason is because it has a very unusual, maybe unique, military doctrine and that is that no military forces are permitted to come under any threat; so if there is any threat at all, they are supposed to react with overwhelming force and that means that in any situation that involves civilians, you know, anything short of total war, the U.S. military simply can't be deployed and in fact isn't if you look at the peacekeeping operations. So that's not a plausible argument, but there are other ones that are plausible and are barely beneath the surface, and that is the very likely concern that an independent judicial inquiry, if it existed, might, as it should, move up the chain of command and that's going to lead it very soon to pretty high places, including the White House. That would be true whether the issue is Indochina or Central America and Panama or Somalia or other exploits.

Well, again, a lot to say about that, but let's turn to the second element of world order: the Universal Declaration of Human rights. Doubtless we will all hear much rhetoric in the coming months about the universality of the high principles that are proclaimed and about the challenge of relativity that's posed by various bad guys around the world. Those charges will be accurate enough, unfortunately, and probably understated, but you are unlikely to hear about a different topic: namely, U.S. adherence to the universal declaration both in action and in doctrine. Well again, and just keep to the doctrine. If you look at the doctrine you find out quickly that the United States is a leader of the relativist camp.

One is unlikely to see headlines about that, but it's pretty clear - the United States dismisses one fundamental component of the universal declaration completely as having no status: that's the component that's concerned with the socio-economic provisions, which have the same status as any others in the universal declaration, but the U.S. doesn't agree. They are a letter to Santa Claus, as Ambassador Jean Kirkpatrick put it. They are preposterous and a dangerous incitement, as they were described by U.S. Ambassador Morris Abrams. He was in fact testifying at part of the discussion by the UN Commission on Human Rights, which was considering a Declaration on the Right of Development which very closely paraphrased the socio-economic conditions of the universal declaration and which, incidentally, the U.S. preceded to veto.

Well again there's more to say, but let's proceed to the third point: the international economic order - the Bretton Woods system, as it's called, and its institutions. That's all over the front pages now with the fears of a global meltdown that might affect privileged folk like us, as well as just the usual victims, so therefore its news. Well, the Bretton Woods system had two basic principles: set up institutions like the World Bank and the IMF, which are called the Bretton Wood institutions - but it had two basic principles which one has to keep in mind, they're important. One principle was to liberalize trade; a goal was to liberalize trade, more free trade. The second principle had to do with capital flow and it was the opposite.

The goal was to regulate capital flow and control it, keep fixed exchange rates, keep capital controls and so on. That was agreed by both the U.S. and the British negotiators - the U.S. main negotiator was Harry Dexter White, the British, John Maynard Keynes - and it expressed a very common conception at the time, which has a lot of plausibility. It's built into the rules of the IMF. There is now an effort, the U.S. is leading an effort, to try to change those rules, but up until now they have been breached many times. The rules of the IMF still authorize countries to regulate capital flow and they prohibit the IMF from giving credits to cover capital flight. Many of you who follow these affairs all know how well that one's been observed, but anyhow it is a rule.

There was thinking behind this, there were reasons. The reasons were in part theory, what some international economists call an incompatibility thesis, which in fact remains the guiding principle of UNCTAD, the main UN Conference on Trade and Development. The theory is that capital flight, so short-term speculative flows which lead to exchange rate fluctuations and so on, that they are going to undermine trade and investment, so they are inconsistent with one another. You can't liberalize both; and recent experience is, consistent with that assumption. The second reason was not a theory, it was a truism. The truism is that free flow of capital definitely undermines democracy in the welfare state, which was at that time far too popular to ignore (it's the mid-20th century).

The basic point is that capital controls allow governments to carry out monetary and tax policies to sustain unemployment incomes, social programs, maintain public goods, without fear of capital flight, which will punish this irrational behavior (irrational in that it's only for the benefit of people, not for the benefit of investors and speculators, and it will be punished by capital flight for obvious reasons). That's the essential point - free flow of capital quickly creates what some international economists call a virtual senate of financial capital which will impose its own social policies by the threat of capital flight, which leads to higher interest rates, economic slowdown, budget cuts for health and education, recession, maybe collapse. It's a powerful weapon.

All of that was articulated quite explicitly in essentially in the words at the time by the U.S.-U.K. negotiators, and it's not particularly controversial. (In fact not controversial at all then or now. If you think it through it's kind of obvious, as it was to them.) And all of that is quite important to keep in mind in looking at the current period because there's a challenge to that in the last 25 years and we see the consequences. (And it's now being re-evaluated because the consequences are even hitting the rich people and that's where we are now.) Well, the Bretton Woods system as formulated that is efforts to liberalize trade and regulate capital that was in place ... to a substantial degree through the first half of this period, the first quarter century after it was established. That's what's sometimes called the golden age of post-war state capitalism - high rates of growth of the economy, of productivity, expansion of the social contract right through the '50s and '60s.

The system was dismantled from the early 1970s. Richard Nixon unilaterally abrogated its basic principles; other major financial centers joined in. By the 1980s capital controls were mostly gone in the rich countries and the smaller economies like South Korea were simply compelled to drop them. That, incidentally, is widely regarded now as a major factor in its recent collapse, alongside of quite extreme market failures in the private sector throughout East and Southeast Asia and of also the west, which was involved in crazed lending. In the light of the recent economic crisis in East Asia, the more serious analysts recognize and insist that the East Asian economic miracle was quite real.

So one of the most important and influential, and Joseph Stiglitz, who is now the chief economist of the World Bank (he was formerly head of council of economic advisors here, and it plays a very important role) he emphasizes in recent World Bank publications and elsewhere that this is post-crisis - that the East Asian economic miracle was not only real but it was in his words an amazing achievement historically without precedent and, furthermore, he points out, based on very significant departures from the official doctrines of the so-called Washington consensus and that it should last, it should thrive, in fact, unless it is destroyed by irrational markets as it could be. Stiglitz points out - remember this is the chief economist of the World Bank in a World Bank publication - that in East Asia the basis for the amazing achievements and the miracle, which has no precedent, is that governments took major responsibility for the promotion of economic growth, abandoning the religion that markets know best, and intervening to enhance technology transfer, relative equality, education, health, along with (he doesn't stress this but he should have) industrial planning and coordination, and in fact strict capital controls until they were forced to relinquish them in the last few years.

Stiglitz also mentions, though he doesn't go into it, that the rich countries, every one of them - from England on through the United States up to the present, every single one of them had followed a somewhat similar path, actually far more so than the World Bank has yet acknowledged. Another big topic can't go into, but an interesting one and again worth keeping in mind. Well what has happened since the Bretton Woods system essentially collapsed in the early 1970s? It did end the golden age of post-war state capitalism. Just focusing on the rich countries, primarily the United

States and Britain, although it happens to others in various degrees in an integrated economy, over the rich countries as a whole, the growth of the economy and the growth of productivity have slowed very markedly. Actually, contrary to what you read, trade also slowed, if you look closely, in the United States specifically and England. Incomes stagnated or declined throughout this period for the great majority of the population; working conditions deteriorated, social services have been significantly cut, infrastructure is in serious danger with very little required public spending, the welfare state has significantly eroded.

There has also been a closely correlated, dramatic increase in incarceration. It's closely correlated because a large part of the society is just becoming superfluous for wealth formation. In an uncivilized society you send out the death squads to kill them; in a civilized society you throw them in jail. Since 1980, when this system really took shape, when it was in place, at that time incarceration rates in the United States were roughly like that of other industrial countries, kind of at the high end but not off the scale, and so crime rates in the United States are not unusually high, contrary to what you read. Again they're sort of toward the high end but not unusual, with one exception: namely, killing with guns. But that's a separate matter that has to do with laws, cultural patterns and so on, it doesn't have anything to do with crime. And that remains the case. In fact, crime rates have declined since 1980, but the incarceration has gone way up. It's a direct reflection of the inequality and the need for social control.

It tripled in the 1980s and it's been rising very fast through the 1990s; it's now five to 10 times as high as other industrial societies. In fact the U.S. is world champion in imprisoning its population, at least among countries where there is any minimally reliable statistics. If you take the prison population into account, that adds another two per cent to the unemployment rate, which places the U.S. squarely in the middle of the European level. Actually, even without that it's not at the bottom, believe it or not, it's about 30 per cent. Of course, you have to decide what you're talking about; if you count in prison labor, which is not trivial, and very good for folks like Boeing Aircraft and AT & T and others (a terrific work force), if you count them in, then of course the unemployment figures change again. Anyhow these two parallel developments have been going on and have integrated.

Throughout the same period profits have soared, particularly in the 1990s. The current jitters on Wall Street have to do with the concern that there may be an end to what for the last couple of years the business press has been calling this stupendous and dazzling and extraordinary growth of profits. They've run out of adjectives and they may now be worried that the facts are ending too. There has been an astronomical increase in capital flows, a huge increase, mostly very short term. About 80 per cent now is estimated to have a round trip — it goes out and back, in a period of a week or less, often hours or even minutes. That means it's virtually unrelated to the real economy, to trade and investment. In fact, current estimates are that about five per cent of the roughly trillion and a half-dollar per day capital flow is related to the real economy. The rest is speculative. If you go back to 1970, the figures were essentially reversed. It was about 90 per cent of a much smaller sum was related to the real economy and maybe 10 per cent was speculative. It's also based very heavily on extensive borrowing; it's highly leveraged, in the jargon, and that's a factor that accelerates something which is often also ... oops... whoever owns these microphones is going to have a problem. Ignore them? OK. Happy to ignore microphones all the time.

This high borrowing, this highly leveraged character of investment, which is something new, incidentally (a lot of things are old but this is new), that's accelerating the irrationality of financial markets. They've become much more volatile and unpredictable; there have been wildly fluctuating exchange rates related to speculative flows and there have been increasing financial crises. The IMF recently did a study of the period 1980-1995, a 15-year period, and it found that about 80 per cent of its roughly 180 members had had one or more banking crises, ranging from significant to quite serious. Again, that wouldn't have surprised Keynes and White or any of the framers of Bretton Woods, or the economists' thinking behind them. In the same period, again in conformity with their thinking, has been an assault, an attack on free markets, a sustained assault on free markets, to quote the head of economic research of the World Trade Organization, in a major technical study. That was led by the Reaganites.

They were talking free markets for the poor but doing something else for the rich. This analyst, Patrick Low, estimates the effect of Reaganite protectionist measures at about three times as high as those of the other industrial countries which were bad enough. Well again, that's what was expected. During the Reagan years, lots of lofty rhetoric but protection was virtually doubled. The public subsidy, which is another violation of free trade principle, was increased, bailouts increased, both for domestic banks and international banks. It happened throughout the world but mostly in the United States - In the United states the goal was to somehow overcome very serious management failures that were leading to a decline of U.S. industry and were a matter of great concern at the time. Those of you who read the business press remember a lot of discussion and concern about the need to reindustrialize America. American industry was collapsing, mostly because of management failures. The Pentagon was called in to fill its traditional role to do something about this problem. (That's actually a role that goes back to the early 19th century before there was a pentagon.)

The pentagon was called in to develop a program under Carter which was greatly extended under Reagan, to design what was called the factory of the future, based on lean production and automation and other developments in which the American management had fallen way behind and then to hand it over to industry as a gift. The purpose was to save central components of the industrial system from mainly Japanese competition, which was wiping it out, and to place them in a position to dominate the emerging technologies and markets of the next era. The Internet and information technology, generally, are rather dramatic examples of this but not the only one. All of this continues under Clinton, alongside the free market rhetoric. Radical interference with free trade is standard when convenient. And it's across the spectrum. So Mexican tomatoes were effectively barred from the U.S. market, as was openly stated, because U.S. con-sumers prefer them and they were undercutting Florida growers, sort of at the other end of the trading spectrum. High tariffs were a couple of months ago introduced on Japanese supercomputers to protect U.S. manufacturers like Cray Enterprises, which is called private enterprise because the profits are privatized. (The markets are public and much of the technology and funding is public as well but the profits are private.)

If you want to see the real meaning of free trade and neo-liberalism in its cruelest form, just take a look at the relation between the richest and the poorest country of the hemisphere - the United States and Haiti. Haiti was forced to liberalize radically as a condition on terminating the terror and torture of the coup regime, which was pretty awful but you didn't have to be there to know it. The cost of liberalization is quite severe. One effect is that Haitian rice production, one of their few potential economic strengths has been seriously harmed and virtually destroyed because it is now competing with US agri-business, which is crazy to begin with, and even crazier when you recognize that 40 per cent of its profits come from government subsidies, thanks to Reaganite contributions to free trade. Recently, the United States has started dumping chicken parts in Haiti, undermining another.

The reason is that American consumers don't like dark meat, so the producers, these big factory farms, have a lot of extra dark meat, so why not dump it on Haiti? We're to wipe out one of the few hopeful enterprises that had developed there. They can't dump it on Canada because Canada has huge tariffs to block that kind of behavior. Haitian tariffs are forced to be, roughly one-fiftieth of what Canada's are, same with the Dominican Republic and Jamaica, but Haiti has to liberalize. Just within the last few days, U.S. steel manufacturers have been demanding that the U.S. government force Japan and Russia to cut back steel imports into the United States; they are particularly worried about Japan because it's high quality steel, which is undercutting them. And probably they will. The U.S. has instruments to do that. Super 301 it's called: you threaten to close off the market to a country and if they don't do it you tell them. And of course Haiti, since it's a free and equal world, Haiti has the same instrument: they could object to U.S. dumping of chicken parts by threatening to close off Haitian markets to U.S. exporters, just as the U.S. can do, so it's all free and equal. Well, that's free trade.

Without going on with that, for the Third World generally, given the relations of force, the post-Bretton Woods era, the last 25 years, have been pretty much a disaster. Some have escaped, mainly by not playing by the rules the way the rich countries have done. Russia is a dramatic example since it returned itself to the traditional Third World role about 10 years ago. Well, there's a standard picture about all of this for the United States. The standard picture is that the United States has a fairy tale economy, that Americans are smug and prosperous in the happy glow of the American boom, there's a fat and happy America enjoying one of the healthiest booms in American history - these are all quotes from front page headlines in the *New York Times*, fairly typical. They all give an example, the same example, up until this summer at least; the example was the stock market, and it indeed is a fairy tale, especially for the top one per cent of households who own about half the stock and other assets and to some extent for the 10 per cent who own most of the rest. Well, what about the next 10 per cent, you know the 80th to the 90th percentile, right below the top ten per cent? What about them? Well for them their net worth has declined in the 1990s for the reason that debt, which is enormous, has increased faster than the growth of stock and other assets.

And it just gets worse as you go down. Eighty per cent of families work a lot more hours just to keep from losing even more ground; they have not yet recovered the levels of 1989, let alone (that's comparable stage of the last business cycle), let alone 1973 - that's when the new economy really began to take hold. All of this is without precedent in American history. It's never happened before. It's the first time that during an economic recovery that these were the consequences: you can't even catch up to where you began for a large majority of the population. As far as economic growth is concerned during this fairy tale boom, it's roughly at the average for the OECD, the rich countries; as far as growth of per capita income is concerned, it's below the OECD average - it's actually roughly like the anemic '70s and '80s and nowhere near the golden age. But it's a fairy tale for some and those are the ones who tell us about it. Those are the Americans who are smug and happy, the rest are some other thing.

The reason for the fairy tale is in fact frankly explained, for example by Alan Greenspan, fed chair. He attributes it to what he calls significant wage restraint and greater worker insecurity. The Clinton administration in its economic report attributes it to salutary changes in labour market institutions, which is a delicate way of saying the same thing. The business world agrees. If you look at the business press, they point out that workers are too intimidated to seek some share in the good times. Just this week Business Week reported studies showing that 60 per cent of workers are very concerned about job security for working people and 30 per cent are somewhat concerned. When 90 per cent of the work forces are insecure, that helps keep profits up and inflation low enough to please the financial institutions, so it's a fairy tale economy. Well, there are a lot of reasons for this. One reason is simply the threat of job transfer if people raise their heads; another is the destruction of unions, which really took off during the Reagan years by straight corporate crime which was authorized by the Reagan administration — again the business press has been clear and frank about this.

These are specific social and economic policies designed to keep things this way; that includes the investor rights agreements. That's a long story in itself, as you should know, or if you don't you should quickly find out. The OECD, the rich countries, are seeking to ram through the Multilateral Agreement on Investments, the sort of super investor rights agreement, in October. (You ought to know it because Canada has been unique in that there has been substantial public opposition to this.) They're planning to do it in October, in secret if they can; they've been trying to do it in secret for a long time. They failed last April and that caused near panic in business circles - it's worth looking at. The Financial Times in London, sort of the world's premier business daily, had an agonized article after they failed about what they called the horde of vigilantes who descended on the OECD countries and the corporate world were totally helpless in the face of this massive assault by Maude Barlow and such, and they had to collapse.

You really have to read it to get a picture of the panic. It also quoted trade diplomats who warned that unless this crisis of democracy is overcome, it may become harder to do deals behind closed doors and submit them for rubber-stamping by parliaments, as in the good old days. Well, that tells you very clearly what it's all about. It's again the hazard-facing - the corporate sector - in the rising political power of the masses, that's been frightening rich and powerful people ever since the first modern democratic upsurge in 17th century England. Well, there's a ton more to say about this but it's getting late. Question: Is this globalized economy really out of control? Well it's very hard to believe that. It's a large majority of the exchanges, the international exchanges, are within what's called the triad - North America, Europe and Japan. These are all areas that have parliamentary institutions, they don't have any fear of military coups, which means what's going on is in principle subject to public policy decisions and can be made in practice so as well. And well beyond that - that's all within existing institutions, assuming existing institutions don't change at all - but that's a pretty strong assumption.

No one should have ever made it in the past, certainly, and there's no particular reason to believe that some magic moment has come. In general, institutions are not self-legitimizing - they've got to legitimize themselves. We live in a world which is largely dominated by unaccountable private tyrannies and they have to justify themselves. They are not automatically self-justifying. When they were created in the United States by radical judicial activism early in this century, conservatives, (who used to exist, they don't anymore except in name) bitterly condemned this change which they saw as a major attack on classical liberal ideas and fundamental theories of human rights. They condemned it actually as a form of communism and a return to feudalism, which was not totally inaccurate.

Anyhow, the institutions are not self-legitimizing. They are internally tyrannical, they are unaccountable to the public, they administer markets through their internal operations and through strategic alliances with alleged competitors, and they are backed by powerful states which provide subsidies and risk protection and bailouts if needed, and so on. And there's a question as to whether those institutional arrangements are necessary and appropriate, a very serious question. It's entirely natural for the doctrinal institutions to try to direct the public attention somewhere else (in fact it would be astonishing if that were not true) to direct attention away from crucial issues and also to try to induce a general mood of hopelessness and despair - what Linda McQuaig in a recent book on Canada, a good book on Canada, calls the Cult of Impotence, she's describing how it works here and to drive people towards individual survival strategies makes a lot of sense to try to do all that.

It's understandable and understanding it can be liberating, as always; it can liberate people to design and follow, if they choose, very different paths. These may well involve, and in my opinion should involve, dissolving centers of unaccountable power, extending democratic arrangements well beyond to central parts of the society from which they are excluded, and may make it possible to address in a serious way the injustice and the needless suffering that defaces contemporary life and to demonstrate that the human species is not a kind of lethal mutation which is destined to destroy itself and much else in a flick of an eye, from an evolutionary point of view. That is not a completely unlikely prospect, in my opinion, under prevailing conditions of social life.

A couple of microphones out there, so anybody who wants to exploit their existence is free to do so. There's a wide range of things that can be done, they're maybe they're interrelated, but on some issues it's pretty clear, at least it's pretty clear, on what ought to be done and in fact not hard even, because it doesn't challenge the structure of institutions. So take, say, the MAI, which, if you're not familiar with it you ought to be, there's plenty of literature about it, especially in Canada. It's what was described by Business Week as the most explosive trade deal you've never heard of, and the whole headline, the whole description is accurate. It is the most explosive trade deal that's ever been crafted; it gives extraordinary rights to corporations. They were given the rights of citizens early in this century, of people, you know, immortal people, super powerful immortal people, which is already an astonishing attack on traditional classical liberal ideals, and the MAI actually gives them the rights of states.

Canadians ought to know about this since Canada has just suffered from it. Canada was sued by a corporation, The Ethyl Corporation, for daring to try to ban a harmful gasoline additive which is banned in most of the world and theoretically not banned in the United States but not used because it's too dangerous. Canada tried to do the same, the Ethyl Corporation sued them under provisions of NAFTA, which is unextended in the MAI - it's really unclear what they mean, corporations are trying to press these to the limit. It's never been possible before for corporations to sue states, but these new arrangements intend to give them the rights of states. They sued Canada for expropriation because it was taking away their enjoyment of their rights by banning this probably poisonous additive. Ethyl Corporation has got a nice record - it's a major corporation set up by DuPont and GM and all those big guys - its major contribution was leaded gasoline. They knew in the early 1920s that it was lethal but they kept it secret and they had good lawyers and they kept things from happening and for about 50 years it was used with horrendous effects. Finally it was banned, at least in the United States, around early 70s, but then it just goes off to the Third World where there are no controls so you can kill anybody you like.

That's the Ethyl Corporation and now they want to import-export MMT into Canada - Canada backed down and paid some indemnity, 13 million dollars or something. There's another case coming along by a hazardous waste disposal company in the United States and there will be more. The idea is to give corporations not only the rights of super powerful immortal persons, which is questionable enough, but even of states, and to undermine democratic options that might be open to citizens - across the board; whether it's things like set asides for minorities or supporting local enterprise or environmental labour rights, you sort of name it and it's there somewhere. It's not put in those words, explicitly, but the intent is to develop a framework which smart lawyers will then fill in with precedents - that's the way it works. So naturally it's got to be done in secret because they know people are going to hate it. And it was kept under a veil of secrecy - it was kept under a veil of secrecy for literally three years of intense negotiations. Secrecy in a funny sense - the business world certainly knew about it and they were right in the middle of it and publishing monographs about it and so on.

The press certainly knew about it but they weren't talking, in the United States Congress was kept in the dark, the public didn't know, it was pretty much the same throughout the industrial world, Canada was a unique exception. Well, anyhow, that was beaten back last April partly because of unexpected public opposition and it's coming up again in October, so in a couple of weeks. And it'll go through if nobody makes a fuss, you know, with long-term effects. Well, OK, it's clear what to do about that, at least - same thing that was done pretty effectively last time around, but more so next time. It'll come back in some other forum you know, like it'll be written into the conditions of the IMF or some secret forum.

And there are a million things like this. We can list them from A-Z - that's what activism is about, trying to deal with those specific cases of threats to society, and justice, suffering, oppression, whatever it may be; all extremely important but short of a further step what about going beyond putting Band-Aids on the cancer? What about the nature of the institutions? Are they in fact legitimate? Well, that's a serious matter. You know you can't just issue proclamations. If you say the organization of society and its domination by unaccountable tyrannies, which is what it is, is improper and unjust, and it is, you have to consider what the alternatives are and how you move towards the alternatives, if you want to. And those are not trivial matters; they require organized popular movements which think things through, which debate, which act, which experiment, which try alternatives, which develop the seeds of the future in the present society, as Bakunin put it a long time ago. And that's a long-term project.

How do you do that? Well, the same way you got rid of kings and slavery and lots of other bad things through history. There's no magic formula. What you do depends on what the conditions are, where you are, what can be done. But it's possible to have a long-term vision about this, and it's in fact one that draws very much from our own tradition, you know, not any foreign borrowings and all that bad stuff. So if you go back to, say, eastern Massachusetts in the mid-19th century without the dubious benefit of radical intellectuals, working class people were running their own newspapers, artisans in Boston and young women coming off the farms who were working in the textile mills were called factory girls and so on, and they're interesting.

They weren't claiming as we do, you know the radicals among us, that corporations have too many rights, they were claiming they don't have any rights. They were not asking them to be more benevolent. They were not asking for the dictators to be more benevolent, they were saying they had no right to be dictators. They were saying that those who work in the mills should own them - simple, and the communities should run them, and so on. It's not an unusual position. Wage labor in the United States wage labor in the mid-19th century was considered not very different from chattel slavery. That goes way back into the classical liberal tradition, so servants were not really considered people because they were working for somebody else. Abraham Lincoln, for example, it was his position. It was northern workers that were sort of their banner in the civil war. The Republican Party, it was its official platform, you can even read about it in New York Times editorials. It's by no means an exotic doctrine; it makes a lot of sense. And it has very deep roots in the enlightenment and way back.

The same is true of inequality. You go back to the origins of western political thought, and literally mean the origins, Aristotle's Politics, it's based on the assumption that a democratic system cannot survive, cannot exist, except under conditions of relative equality. He gives good reasons for this. Nothing novel or exotic about this. The same assumption was made by people like Adam Smith. If you read Adam Smith carefully and he was pre-capitalist, remember, and believe anti-capitalist in spirit, but if you look at his argument for markets, it was a kind of a nuanced argument, he wasn't all that much in favor of them, contrary to what's claimed. But when you look at the argument for markets, it was based on a principle: the principle was that under conditions of perfect liberty, markets ought to lead to perfect equality; under somewhat impaired liberty, they'll lead to, somewhat, a degree of inequality. And equality was taken as an obvious desideratum, you know, a good thing.

He wasn't thinking about democracies, he was thinking in other terms. These are important ideas. They have to be revived, brought back into our mode of thinking, our cultural tradition, the focus of our activism and the planning for how to change things. And it's no simple business. It wasn't easy to get rid of kings, either. First of all, this has been a pretty horrible century, one of the worst centuries of human history in terms of humanly created disasters and catastrophes, many of which but not all, but some of the worst of them, come from the peaks of western civilization. But in many other respects, it's a lot better than it was. If you look realistically over time, you know it's kind of hard to say when you see the ugliness around you, but if you look realistically over time, things are improving. Lots of things that were considered perfectly normal and natural say a century ago would be considered outlandishly outrageous today; nobody could even conceive of them. In fact that's even true of the last 20 or 30 years - for many of us our own lifetimes. Things have really changed a lot.

And we know how they've changed - not by sitting around and talking about it. So let's take the last 30 years.

Compare Ronald Reagan and John F. Kennedy. Reagan tried, well, Reagan's advisors, he was probably sleeping, but his advisors basically used Kennedy as their model, more or less, you could just sort of see it in detail. As soon as the Reagan administration came in, it tried to organize a major attack in Central America where all kind of things were going on that they didn't like, like the Catholic church was - there was no clash civilizations then - the catholic church was the main enemy. They really wanted to do in Central America what Kennedy had done in South Vietnam in 1961 and 62 when he basically attacked South Vietnam, you know, sent the U.S. Air Force to start bombing civilians, use napalm, drive people into concentration camps and so on. It was South Vietnam; that was the main target of the U.S. attack. Reagan tried to duplicate that, same mechanisms, same white papers, everything else.

It was a total collapse. After a couple of months of trying they had to back off and the reason is because enormous, unanticipated popular objections were coming from the church, from human rights groups, from everybody. And they had to back off because it was going to threaten other objectives. They actually called the press off and told them to stop the campaign. Kennedy didn't have to worry about that. When he sent the U.S. Air Force to bomb South Vietnam, it was known; you could read it in the New York Times, but nobody cared. In fact people cared so little that the whole era has disappeared from history. Try to find a textbook or even a scholarly book which talks about when the U.S. attacked South Vietnam - we know when the Russians invaded Afghanistan, but we don't know when the U.S. attacked South Vietnam.

In fact, ask educated people, your friends and teachers and so on, to see if they can give you the date of when that took place, and they won't even know what you're talking about. There was no such event in official history. There was such an event in real history, but since nobody cared about it, and if the president wants to go bomb some other country, who cares, it kind of disappeared in to the mist and what was left was the propaganda. Couldn't do that in the 1980s - in fact it was totally different. The popular reaction in the United States to the Central America wars were completely different from in the '60s and much more powerful, again contrary to what people say.

So in the 1960s it never occurred to anybody to go live in a Vietnamese village because maybe that would cut back state terrorism by US clients. Many, many people did that in the 80s and people from the heartland, Midwest rural areas, actually conservative Christians, sometimes fundamentalist Christians. These are things that are completely unheard of in the '60s. And the same is true on a host of other issues. Think about women's rights, or respect for other cultures or environmental issues and so on. They barely existed in the '60s. There was a big change in just 30 years and it's a much more civilized society in many ways. That's not to say that a lot of rotten things haven't been happening - they have. In fact a lot of the things that have been describing in the last 25 years, in my opinion at least, are a pretty conscious reaction to that, an effort to stem the tide, and it's partly worked but not in attitudes. It hasn't worked there.

Well, all of that's important and it shows in a very brief moment what you can achieve, and a lot of it was led by young people, incidentally, so one should feel no limits on what could be achieved. And if you look over a longer stretch of history, yeah, that's true. So take what's maybe one of the most civilized countries in the world today, say Norway. Norway has very humane, by comparative standards, norms of behaviour like treatment of prisoners. But take a look at a book by one of the world's leading criminologists, Nils Christie, who is Norwegian. He reviews the history of incarceration in Norway, and he points out it went up pretty sharply in the early 19th - this is from memory, but something like this - it went up pretty sharply in the early 19th century and he points out that the reason it went up is because the modes of punishment changed.

So before that, if somebody robbed a store, what you did is you'd drive a stake though his hand. OK, so when you did that you didn't need jails...well, you can't even talk about it now. You go back not too far before that in England and people were being drawn and quartered. You don't have to go back very far in history to find things so outlandish you can't even conceive of them. In the 19th century, well-known medical researchers in the United States were carrying out experiments which make you think of Mengele; so a good deal of gynecological surgery was developed apparently by respected doctors who were experimenting on slave women and Irish women, who weren't considered much different. You know repeated experiments until they figured out how to do it right and that sort of thing. That's inconceivable; nowadays that's Mengele, you know, but then it was maybe not very nice, but not all that crazy.

There's good technical literature on the development of information technology and computers and the internet and so on, and it doesn't look, from my reading at least and some experience with it, it doesn't look as if that was a major factor, although it was indeed used very fast for that. So the telecommunications revolution is a substantial part of what has led to this very radical change in the way speculative capital zooms around the world instantaneously, undermining currencies, distorting trade, and so on. Yes, that technology has certainly been used for that. So you can get the whole content of Wall Street resources and stick them in the Japanese stock market because they're 12 hours different, than using it all the time. You couldn't have round trips for capital movement of an hour or even a week if you didn't have fancy technology; and you couldn't have all this highly leveraged lending with sophisticated derivatives and all that crazy business. In fact a measure of it, if you want to see it at work, at MIT, you know, sort of a high class science, engineering school, every year at graduation, corporate recruiters come around and pick up the smart guys who are getting their PhDs.

If you're teaching music at MIT, you're getting paid by the system, basically, the rest is bookkeeping. And that's true since the 1940s and it was pretty conscious. So you go back to the business press in the 1940s and they made it very clear that high-tech industry, cannot survive in a competitive free enterprise economy, and Business Week added, government has to be the saviour. They were specifically talking about the aeronautical industry but the lesson was intended for high-tech generally, because they just need huge public subsidies. That's why the internet was developed, to take a recent case, within the military system, since the 1960s, then taken over by the National Science Foundation, public, and just two or three years ago handed over to private corporations so that Bill Gates and so on can make money from it. Gates at least is honest about it. He attributes his success to the ability to embrace and enhance the ideas of others, usually ideas coming out of the public sector or funded by the public sector. And the same is true pretty much across the board. That's the way the economy works. Take a look at any dynamic part of the economy and you find that it works that way.

Now of course it's applied and it's applied in ways which weren't anticipated, like when DRPA, the Defence Research Project Agency, which initiated the internet and had most of the ideas and so on, when they were developing all this stuff, they did not have in mind that sooner or later it would get in the hands of big corporations who would try to use it for a home shopping service to marginalize people and turn them into passive consumers and so on and didn't have that in mind, but yeah, surely that's what they will try to do. They certainly didn't have it in mind that it would be used to undermine the MAI by getting around the constraints of the media - it was used for that too. So things have all kinds of applications and consequences, but they're basically developed just because you need it for the technology.

Same reason why, when during that period of management failures, the defence department and military in the United States was called on to create the factory of the future. And that goes way back. What's called the American system of manufacturing, which sort of amazed the world in the mid-19th century, based on replaceable parts and mass production, all this kind of stuff? A lot of that came straight out of the Springfield armoury. It was developed for military technology then adapted to production. It's hard to find anything in the modern economy that didn't more or less work like that. It's not always the military. That's what Stiglitz is talking about, chief economist of the World Bank, when he talks about the fact that the path that the East Asian miracle is following is not all that foreign to us, actually much more so then he recognizes.

3

Role of Technology in Society

In the past, our scientific methods and institutions have tended to emphasize the study of individual natural processes rather than systems, analysis more than synthesis, and understanding nature more than predicting its behaviour. And in many instances, science has focused on short-term, small-scale problems, often in monodisciplinary mode, rather than on long-term, large-scale or integrated problems. While these approaches and perspectives have built up a considerable base of knowledge and led to a vast portfolio of useful technologies, especially in the 20th century, many of the problems now facing humankind can be solved only if we approach science more holistically. Greater effort is needed to *understand* integrated natural systems on multiple time and space scales.

Scientific findings must also be applied at the right scales. The impact of technological interventions on individual people, communities and the environment must also be carefully considered. To do this, science needs to become more multidisciplinary and its practitioners should continue to promote cooperation and integration between the social and natural sciences. A holistic approach also demands that science draw on the contributions of the humanities (such as history and philosophy), local knowledge systems, aboriginal wisdom, and the wide variety of cultural values. The influence of science on people's lives is growing. While recent benefits to humanity are unparalleled in the history of the human species, in some instances the impact has been harmful or the long-term effects give causes for serious concerns.

A considerable measure of public mistrust of science and fear of technology exists today. In part, this stems from the belief by some individuals and communities that they will be the ones to suffer the indirect negative consequences of technical innovations introduced to benefit only a privileged minority. The power of science to bring about change places a duty on scientists to proceed with great caution both in what they do and what they say. Scientists should reflect on the social consequences of the technological applications or dissemination of partial information of their work and explain to the public and policy makers alike the degree of scientific uncertainty or incompleteness in their findings. At the same time, though, they should not hesitate to fully exploit the predictive power of science, duly qualified, to help people cope with environmental change, especially in cases of direct threats like natural disasters or water shortages.

The current trend toward privatization in many countries is influencing the focus and practice of science. While in some instances the net result may be to increase research capacity and knowledge in selected areas, there is major concern that the trend may be undermining public-sector science, especially fundamental research and efforts to solve socially important problems of no interest to commercial enterprises. Patent protection of private intellectual property, for example, makes the job of public research more difficult. There is also concern over the social implications of private ownership and control of technology, and its effect on broad public scientific literacy, and on options for public choice. Another major trend shaping science is globalization.

The end of the Cold War, growing technology demand from emerging economies, world recognition of the interconnectedness of the planet's biophysical systems and improved communications, especially via the Internet – all these forces are boosting cross-border scientific cooperation and information exchange between individual researchers, institutions and governments. However, much of the expansion is occurring in just a handful of scientifically advanced countries. For science to be truly global, more effort is needed to ensure all countries, rich and poor, and a wide range of world cultures are included in collaborative research and technology transfer. This is especially important in areas like global climate change which will affect, sooner or later, all human beings. With the right policies in place, joint scientific work in critical areas such as the Arctic, for example, could serve as a model for other types of global cooperation.

A major challenge for global science is to find institutional arrangements conducive to success. The proliferation of international networks and programs, the so-called "acronym jungle", reflects a rather ad hoc approach, necessitated in part by the narrowness of purposes of established scientific institutions and the lack of strategic, integrated support by national governments in areas like global change or international aid. What is needed is the formation of true international partnerships that allow scientists in different disciplines and countries to fully support each other's aims and share resources and management duties to mutual advantage.

Scientists and scientific institutions should:

- Promote multidisciplinary approaches to research, encourage cooperation between the social and natural sciences, and draw lessons from the humanities, local knowledge systems and aboriginal wisdom;
- Encourage a holistic approach to problem solving that takes into account a realistic range of socioeconomic conditions and effects, as well as multiple time and space scales, where appropriate;

- Carefully explain the implications and the inherent limitations of their research findings to the public;
- Fully exploit the predictive power of science to serve social needs with candid awareness of the limitations of scientific predictions;
- Promote the inclusion of scientists from resource-poor countries in international cooperative projects and maximize their access to information and technology;
- Encourage the creation of science-coordination mechanisms at the highest level of the United Nations, fully involving the governments of all countries, as a way to promote integrated responses to global problems.

Communication and Education

Within the general public, there is certain measure of mistrust and even fear of science and technology (S&T). Some is based on public experience, but much is the consequence of a significant communications gap between scientists and society. Many reasons are advanced for these attitudes: public ignorance or misunderstanding of science, inaccurate or biased media coverage, uneven distribution of the costs and benefits of science among different sub-groups in society, lack of public control over the applications of S&T, and the inability of some scientists to communicate ideas in plain language. The issue of nuclear waste disposal is one example of how the gap between scientific findings (which, in this case, suggest that safe disposal technologies exist that are at least as safe as other industrial risks accepted by society) and public opinion and behaviour (continuing opposition to the use of such technologies) may sometimes appear intractable, that is, not amenable to solution simply through improved communication or further technical research.

Good scientific communication via the mass media is especially important in those areas directly and strongly affecting people's lives — for example, before, during and after natural disasters such as storms, volcanic eruptions and earthquakes, as well as in the general area of global change or depletion of natural resources. In communicating their ideas, scientists should make clear the limitations of their predictions and other pronouncements. But they should not shy away from public pronouncements just because their messages contradict public wishes or expectations; indeed, they should be prepared for negative reactions in those instances, and carefully explain the basis for their scientific conclusions or opinions.

Apart from communication by the mass media which is largely unidirectional, communication in the sense of an ongoing dialogue between scientists, the public, and policy-makers is also important. This may take many forms: public policy consultations and review committees, science fairs, open houses, and public information services provided by universities, research institutes and private companies. As the demand for transparency and accountability in science grows, communication of this type as well as public participation in decision making about the applications of S&T becomes imperative. Unfortunately, resources for such dialogue are lacking not only among scientific institutions but among those groups in society who have a particular stake in scientific developments and therefore something to gain through contact with scientists. Increasing privitization of scientific activity also discourages open communication of scientific findings and uncertainties.

Science education, particularly training in multidisciplinary and team approaches to research, is also in need of reinforcement. Many science education programs still focus on individual student assignments and individual evaluation, whereas the trend in both the public and private sector is toward team work, and the needs of society are increasingly met by the concerted efforts of many areas of investigation. Science, if it is to appeal strongly to youth, also needs to be demystified by educators that is, presented in an attractive, stimulating fashion, with the abstractions of theory strongly linked to everyday life. Furthermore, students need to be more fully involved in public discussion of science and its applications. Not only are they the ones who will be most affected by the current direction of science, they are also the scientists and policy makers of tomorrow.

Recommendations:

- To improve the quality of science journalism, the mass media should engage more journalists with scientific training. At the same time, the mass media and specialized educators should be enlisted to help train scientists or their spokespersons in the fundamentals of public communication and to familiarize them with the expectations and operating parameters of the mass media.
- The concept of scientific clearing houses services to help journalists interpret scientific data, decipher technical language, and distinguish scientifically credible claims from unsubstantiated ones should be promoted. UNESCO national commissions should also consider setting up scientific information services aimed at improving the quality and quantity of science stories in the media and ensuring that differing viewpoints are presented.
- Science community partnerships for example, between research institutes, private firms, the media, and governments—are an effective and practical way to share the costs of communicating science to the public. These should be encouraged.
- Educational authorities should encourage teamwork training and multidisciplinary approaches to science education. They should also attempt to demystify science to make it attractive to a larger proportion of students. University and private-sector experience with team-oriented research should be documented and analyzed with a view to identifying the best current practices in North America.

North-South Issues

Science in the developing world differs from that in the industrialized world in three main ways: budgets are much smaller, research agendas are different because the socioeconomic and biophysical problems to be solved are different, and there is a lower level of access to and public understanding of scientific information and technology. The North-South knowledge gap is viewed by some as the most pressing social and economic aspects of modern science. Many developing countries have well-qualified scientists but often they are few in number and lack the resources and political support needed to solve complex problems or to apply their knowledge to national issues.

In Mexico, where agriculture remains an important part of the national economy, scientific work related to food production and food security is complicated by a web of social problems such as rural poverty, social discrimination against peasants, migration to cities because of changes in land use, weak transportation and marketing services, and lack of farmer access to credit. In the area of health, too, the problems of developing countries are much different than those of developed countries. Chagas' disease and schistosomiasis, for example, are endemic in many developing nations, yet they receive very little attention by health scientists and pharmaceutical firms in industrialized countries.

While there are number of North-South cooperative programs to support science in developing countries and improve technology transfer, much more should be done. Water management, tropical disease research, and energy-efficiency technology were identified as areas where the current co-operative programs are weak, but in which the industrialized countries can provide valuable assistance to developing countries. In the case of international research on large-scale problems like global change, most developing countries are unable to contribute to those scientific components requiring sophisticated research facilities and technologies. However, there are other effective but inexpensive ways for them to participate, such as regional monitoring and carrying out studies of local conditions and effects.

It was suggested, for example, that Mexico could contribute to research on climate change by carrying out, at very low cost, epidemiological studies of a possible link between urban air quality and recently observed seasonal increases in cardiovascular disease and pregnancy-related hypertension. ICSU has an important role in ensuring that developing countries are involved in global change studies on imaginative but affordable and practical ways. Another symptom of the North-South science gap is the inequitable distribution of profits generated by new technologies and products based on plant genetic resources obtained from developing countries.

Recommendations:

- Efforts should be stepped up to give developing countries better access to scientific expertise, information and technology, especially in the areas of disaster relief, health, energy, and water management. In particular, the scientific and technical know-how of military organizations should be harnessed to monitor and alleviate the effects of disasters around the world.
- Measures are needed to systematically involve all countries in research on global change. Developing countries' scientific knowledge of local conditions and effects should be harnessed in the worldwide effort to understand, predict and adapt to global change and the growing understanding of changes in climate, water, and soil incorporated in international assistance programmes.
- Countries and communities should be fairly compensated for their contribution of plant genetic resources that lead to commercially profitable technologies.

- As a priority, science should address the basic needs of the sick and disadvantaged in the poorest countries.

Economics vs Sustainable Development

Science today seems caught in a cross-fire between two opposing world views. On the one hand, science is a major tool of the ideology currently driving the world economy, namely that of the free market system, continual growth and the pursuit of personal wealth. On the other hand, science is increasingly being called on to produce knowledge and technology that promote environmentally sustainable, people-oriented development and long-term management of resources.

The world economy continues to rely heavily on cheap oil, a non-renewable resource and major contributor of greenhouse gases. Fossil fuels - oil, coal, natural gas - will continue to power world industry for several decades. The fact that they will do so despite the availability of technically feasible alternative "green" energy technologies brings the dilemma into sharp relief. Examples of the conflict between current economic forces and the need for sustainable development can be found in many other domains as well. The imposition of structural adjustment policies by international financial institutions, for example, has forced some countries to reorient agricultural research and production to focus on cash crops that generate foreign currency rather than food crops for local consumption. In some cases, such policies have put food security and the continued production of the land in jeopardy, created enormous personal hardship for citizens, and led to social unrest.

Free trade arrangements, too, may pose a threat to some of the underlying components of sustainable development, affecting biodiversity, community self-reliance, and local knowledge systems. In some cases, the elimination of trade barriers between countries has led farmers to abandon the cultivation of traditional crop varieties that were well adapted to local conditions and tastes, in favour of imported varieties that may respond better to newly expanded markets. Deregulation and privatization are two trends aimed at improving commercial competitiveness, and stimulating economic growth. Yet in some sectors such as energy production and food it is becoming clear that these trends cannot be reconciled with the requirement imposed by sustainable development that hidden environmental and social costs of economic production — that is, costs born by present or future society but not normally reflected in prices of goods and services like energy, be taken into account.

In the past, developments in the energy field have had more to do with the protection of vested economic interests than with concern for the public good or environmental conservation. The prospect of that approach being perpetuated is a major concern for the future of energy science, since fossil fuels are a finite resource and a major contributor of greenhouse gases and research or energy alternatives is handicapped.

Recommendations:

- Policy makers must accept that, for certain key areas like energy development, decisions must not be based only on political expediency — such as the prospect of short-term economic benefits and job creation. To do so denigrates the role of forward-thinking research and development (R&D) and undermines long-term social development. Rather, what is needed is a vision of the world that looks "seven generations" ahead, in the manner of the holistic philosophies of North American aboriginal people.
- Public debate on the dangers of "consumptive" lifestyles typical of the industrialized countries needs to be reactivated. If everyone on the planet lived as many North Americans do, we would need the resources of "seven Planet Earths". As this is clearly impossible, the implications of inevitable major changes soon to come should be openly discussed at all levels of society.

- Scientists need to cultivate a new vision of science – one that promotes the development of sustainable "closed" systems of production and consumption, which are compatible with the recycling behaviour and equilibrium of natural systems.
- Agencies that provide research grants should be broader in their terms of reference and more neutral and flexible so that scientists are not continually pushed to find short-term solutions when long-term ones are needed. In some countries, the allocation of research funds is controlled by small powerful groups who engage in favouritism for their own personal gain or prestige. Governments should ensure that systems for evaluating and funding project proposals are fair, objective, and transparent.

Science Policy and Ethics

Scientific advances are never, in themselves, a guarantee of social benefit. Technology has to be treated as a servant of society, not a master. Increasing commercial productivity, while at the same time necessary, unemployment and poverty is not a socially acceptable solution. Science must be fully integrated with broad societal needs, but this tenet is not yet fully accepted. One reason for public mistrust of science is that ordinary people feel they will sometimes end up being the ones to suffer the costs of technological innovation. It was suggested repeatedly at the North American meeting that the time has come to introduce an international code of ethical conduct for scientists to ensure that science is directed for the public good.

Scientists in their daily work are sometimes isolated from mainstream society, making it difficult for them to be clearly aware of public needs. Conversely, policy makers, in need of sometimes urgent advice on technical matters, sometimes urgent, may be unaware of the scientific expertise residing under their very noses. Society has much to gain by the proactive involvement of scientists in policy making.

Medical biotechnology is a leading-edge area of science in which the pace of progress is perhaps faster than society's capacity to deal with the ethical and social implications. Genetic research, while offering major benefits for disease diagnosis and treatment, also poses serious questions about the nature and sanctity of human life and the protection of human rights. The possibility that genetic technology could be commandeered by powerful groups to pursue goals in their own interests but which may be socially destructive or discriminatory is not to be considered lightly. It is an issue of particular importance to disabled persons. Greater dialogue between scientists, policy makers and the public, especially those groups disproportionately affected by technological developments, is clearly needed.

A major concern is that recent advances in health sciences will lead to the "genetification of medicine", that is, a trend toward understanding and explaining human beings and human health largely in terms of genes and their interactions. A worry here is that the role of environmental and social factors will increasingly receive insufficient attention, leading to a one-dimensional view of diseases and disabilities. A further ethical issue for science is what has been referred to as the "commodification" of basic human needs such as food, shelter, clothing, fuel and health services.

In many countries, many of these items have traditionally been supplied through non-monetary social support structures, often family-based. As cash economies and government welfare programmes increasingly treat these necessities of life simply as commodities to be bought and sold, there is a serious risk that technological innovations, stimulated by scientists working within a commercial framework, will be exploited mainly by well-to-do minorities, with little or no benefit to the poor. The potential of science to improve human social conditions in non-material ways needs much more attention.

Recommendations:

- The gaining of scientific knowledge must not be assumed to lead automatically to direct commercial policy exploitation of that knowledge. Often the knowledge is of greatest benefit if it increases public understanding and awareness. Scientists cannot always control the application of their findings. However, they have a responsibility to engage in public dialogue about the implications of scientific findings and to help distinguish between socially beneficial and socially harmful applications.

- Action is needed at the international level to protect the human species from human-induced genetic alteration and to ensure that technological applications in the fields of human genetics are ethically and socially sound. Review committees at the institutional and national levels, such as those that examine and appraise research projects, can help focus attention on key ethical and safety issues. However, stronger and higher-level mechanisms for decision-making and enforcement in this area of science are also needed. UNESCO has an important role to play in this regard.
- Scientists should be more proactive in policy making. This could be done by promoting, among governments around the world, the concept of "science/policy contracts". These agreements would recognize the value of scientific advice, but also make clear that such advice is but one ingredient in decision-making and not necessarily the overriding one. Such contracts should set clear performance standards by which the inputs of scientists can be evaluated.
- The world scientific community should consider adopting an international code of ethical conduct for scientists, similar to the Hippocratic Oath taken by physicians. This code would apply a similar principle of measurability to scientific behaviour that scientists so cherish in their day-to-day pursuit of knowledge.
- (In a commentary subsequent to the workshop, one participant suggested that the Engineer's Pledge, which undoubtedly has influenced the ethical conduct of

professional engineers in several countries, could also be a model for principles of conduct of science in general, adapted to express consideration for all of humankind, ecological integrity, and long-term consequences).

Integrating Issues-Science and Society

Advances in science and its resulting technologies, such as global communication, satellite images of Earth, together with the popular fascination with dinosaurs etc., have irrevocably expanded the space and time scales with which people at many levels of society now view their world. Science is largely responsible for a growing public awareness that people share the planet with all other living creatures, that the environment which supports all life is subject to change, and that human activities are presently changing this environment and threaten to change it seriously. In the past two centuries, science has been used mainly as a tool for economic expansion and military power for the wealthier segments of the human race.

It is now clear that the current consumption of natural resources and increasing stresses on the regional and local environment cannot continue indefinitely without breakdown of the natural support systems that make present civilizations possible. Science, which helped to bring about this situation, now has an over-riding responsibility to help societies make a transition from an obsession with growth to achievement of a dynamically stable and sustainable ecological and economic system. In this transition, an alliance between modern technical science and the holistic wisdom from indigenous societies and philosophers from all cultures can be very important. In the coming century, the rate of change of natural and human conditions and issues can be expected to continue to accelerate.

Scientists have an increasing obligation to become involved with policy-makers and the public in finding and implementing solutions or means of adaptation to issues that are both local and world-wide, such as reconciling the present competitive profit motive with the common good; providing for contributions from and benefits to marginalized elements of society and minority cultures; justifying current expenditures to prevent costs or damages to future generations; rewarding collective rather than individual efforts. The role of science in society and governance has never been more important.

Constructivism

Technology has always impacted education; the printing press allowed textbooks to be developed, and the replacement of slates and chalk by pencil and paper permitted a permanent record of one's writing to be preserved. In the late 1950s and 1960s television was utilized as a means of teaching large groups of students, albeit ineffectively. Today, a new wave of technology is beginning to cause repercussions in schools that will forever change how students are taught. Is this change a positive one? To gain insight into this dilemma, one must first look at how children learn. Many theories of learning have been proposed in the last century. Until recently, behavioral psychology has influenced education in this country to such a startling degree that it had a virtual stranglehold on how textbooks were defined and how teachers planned and implemented lessons. To date most statewide testing programs reflect this philosophy.

Of late, however, the constructivist theory of learning has, again, come to the forefront. Constructivism isn't new; John Dewey advocated it at the turn of the century. Maria Montessori based her educational model on constructivism as did Joseph Bruner, and more recently, Vygotsky. Constructivism as a theory of learning is juxtaposed to behavioral theory. Constructivism is child-centered; it "proposes that learning environments should support multiple perspectives or interpretations of reality, knowledge construction, context-rich, experience-based activities".

Constructivism focuses on knowledge construction, not knowledge reproduction.

It is a belief that one constructs knowledge from one's experiences, mental structures, and beliefs that are used to interpret objects and events. Jonassen (1991) stated, "The mind is instrumental and essential in interpreting events, objects, and perspectives on the base that is personal and individualistic". Our view of the external world differs from others because of our unique set of experiences. Senge (1995) put it this way: "We don't describe the world we see; we see the world we can describe."

An important component of constructivist theory is to focus a child's education on authentic tasks. These are tasks which have "real-world relevance and utility, that integrate those tasks across the curriculum, that provide appropriate levels of difficulty or involvement". It would be impossible for us all to become masters of all content areas, so "instruction is anchored in some meaningful, real-world context". According to constructivist theory, children learn whole to part, not incrementally.

The ideas and interests of children drive the learning process. Teachers are flexible; sometimes they are the giver of knowledge, but often are the facilitator. Holden (1994) states, that learning must become the constant and time the variable in a constructivist setting. Dwyer (1991) asserts that this approach is child centered rather than curriculum centered, while Bagley and Hunter (1992) state that learning becomes a dynamic process. Bagley and Hunter (1992) go on to say that active learning leads to greater retention and higher level thinking. And as knowledge continues to double every two years, and since it also has a shelf life, students must learn to access information; there is now far too much information to memorize.

Acceptance in Public Schools

What problems exist for constructivist classrooms? The general public is often suspicious of educational practices which differ from what they experienced. State-wide assessment is not in alignment. Students are required to perform on standardized tests which do not assess what they are learning. The standard report card with letter grades also causes problems for both teachers and students. And the structure of the school day causes most classes to be fragmented. The clock rules!

The Role of Emerging Technologies

Education is being partially transformed by new technologies. At one time students could learn a small, but fixed body of knowledge. However, today, the enormous amount of available information, coupled with the fact that the amount of knowledge in the world continues to double at an increasingly quick rate, requires a transformative approach to education. It is imperative that the student of today learns how to be an information manager, rather than in information regurgitator. In a technology-rich environment one must remember that the educational focus is on learning and instructional goals instead of the technology itself, because technologies are merely tools or vehicles for delivering instruction. It is not what equipment is used, but how the equipment is used which makes it relevant to a constructivist classroom.

Studies show that in technology rich classrooms there are many observable changes:

1. There is a shift from whole class to small group instruction.
2. Coaching occurs rather than lecture and recitation.
3. Teachers work with weaker students more often rather than focusing attention on brighter students as in traditional settings.
4. Students are more actively engaged.
5. Students become more cooperative and less competitive.
6. Students learn different things instead of all students learning the same thing.

7. There is an integration of both visual and verbal thinking instead of the primacy of verbal thinking.

In one study, student self-esteem and motivation in a technology-rich environment was measured and found to be strong. In addition, student attendance was up and discipline problems were reduced. Students were also coming in on their own time—lunch, recess, after school—to work on their projects. Students shifted from being competitive to collaborating on projects.

Technology Meets Constructivism

Collins (1991) states, "So, inadvertently, technology seems to be coming down on the side of constructivists, who have been trying—unsuccessfully to date—to change the prevailing societal view of education". Why? Because computers undermine the didactic, lecture methodology, and, instead promote the student as a self-directed learner. "And just as a change in practices with respect to racial integration led eventually to a change in racial attitudes, so a change in practices will slowly lead to a change in the educational beliefs of society. Using computers entails active learning, and this change in practice will eventually foster a shift in society's beliefs toward a more constructivist view of education".

According to Mann (1994), the use of new technologies in an educational setting has caused the theory of learning, constructivism, to receive new attention. Students in these settings become empowered by gaining access to real data and work on authentic problems. Often, roles are reversed as teacher and student learn from one another. Strommen and Lincoln (1992) make the point that it is not what equipment is used, but how it is used that makes the difference. "The key to success lies in finding the appropriate points for integrating technology into a new pedagogical practice, so that it supports the deeper, more reflective self-directed activity children must use if they are to be competent adults in the future".

In other words, computers and other technology should not be viewed as "add ones," but as tools which are an integral part of a child's learning experience. It has been suggested by LeBaron and Bragg (1994) that the role of technology in education is so important, that it will force the issue of didactic versus constructivist teaching. Teachers will no longer have a choice but will be compelled to use a constructivist approach in a technology-rich environment. How can technology be appropriately used, and how can it be misused?

There are many software packages available which are little more than electronic worksheets focusing on skill and drill. Compoy (1992) states, "To promote the wholesale use of technology-based systems for the teaching of mechanical skills, means that schools would settle for the lowest level of instruction to the exclusion or extinction of higher level skills. If the present system is failing to produce the type of educated students that the nation needs, then automating those same processes will not change the educational outcomes. Having students learn superficial information faster will not improve education".

Technology, can, however, be used effectively in many ways, some of which are described in this chapter. Barr's (1990) five goals essential to meaningful educational reform apply here. He states that learning should be more:

1. Independent
2. Individualized
3. Interactive
4. Interdisciplinary, and
5. Intuitive.

Appropriate use of technology in the classroom naturally meets these five goals. Barr (1990) states that resources in a typical school library are usually quite limited and "often impose artificial restrictions which limit the potential for genuine exploration and discovery". In addition, with current knowledge having a shelf life of approximately one and a half years, many library materials are out of date or incomplete. Available resources via technology now include electronic data bases with current information and other independent research tools available what used to be accessible by only the most sophisticated research environments. Barr (1990) goes on to say, "If we wish to prepare students for life-long learning, we must begin to introduce them to the tools which they will use in the careers they pursue after their formal education is completed".

Perhaps the most useful pieces of available technology for students are hypertext and hypermedia which allow students to browse information in a nonlinear fashion. These data bases contain hyperlinks which give the decision making power as to what to explore next. This type of interactive learning also allows the student to create his or her own nonlinear data bases! "Interactive learning in this context means learning in which inquiry, feedback and ongoing collaboration play important roles". One commercially available hypermedia application is Hyper-card. In a Hypercard environment the focus is on single, complex projects which require students to "conceive of and execute an entire program which presents them with a rich, open-ended, self-directed tasks in which they can explore the various procedures and how they interact rather than simply learn them in isolation from one another, in rote fashion".

Students select their own project content, but in reality actively collaborate with other students. Thus, without prompting, the students' projects become a group effort as they tutor each other in new procedures and review each other's work and make suggestions for improvement or revision. In addition, students use more resources, seem to enjoy learning more and generate higher level reasoning strategies as well as demonstrating a greater diversity of ideas and procedures.

Much work has been ongoing, particularly at the secondary level. Charles Jervis's students at Auburn High

School are creating HTML documents so that their work can be posted on the World Wide Web for peer review. Suzan Mauney and Donna Swenson at Blacksburg Middle School work with eighth graders who are creating a multimedia magazine which will be highlighted on a web page. Mark Freeman at Blacksburg High School has posted his physics students' work on the web. Each of Mark's students researched a topic from opposing views. Both views are posted on Blacksburg High's web page. A form for feedback is provided.

This allows each student to get feedback from members of a real world audience who vote for the view with which they agree, based on the persuasiveness of the student's argument. As a facilitator for the NSF planning grant, it has been part of my job to assist with these projects upon request. In addition, I have created web pages for three schools and regularly post student work for real-world feedback. I have witnessed first hand the positive impact this has on students.

Resistance to Technology

Overall, however, problems do continue to exist. Lack of funding and no clear vision keep systemic change from occurring as rapidly as the evolution in technology. Most elementary schools in Montgomery County are lucky to have old Apple IIe computers which are out of date and cannot support software such as Netscape or Blue Skies. When new technology is purchased, it is generally bought for secondary schools. Donations made to the schools are not new technology, but old computer equipment incapable of handling the rich world of the internet, particularly the World Wide Web. In addition, too many teachers are unaware of much of what is available, so awareness sessions must be held.

While Montgomery County has a technology plan, budget cuts have reduced the possibility of its implementation. This plan also does not address curricular issues. What is greatly needed is a major paradigm shift. To prepare for this paradigm shift, teachers need to be trained and to think about why they do what they do. Acculturation in the schools must take place, but this is a slow process. Elementary teachers often use a constructivist approach, but most secondary teachers continue to teach in a didactic manner. Thus, even though students come to us from a technology-enriched environment where they control information flow, they are expected to fit into an educational institution unchanged by the technology which has swept through society.

Another problem concerns student teachers who regularly come into classrooms having never turned on a computer! Strommen and Lincoln (1992) chastise teacher training programs by stating that because of a lack of training in the uses of technology, student teachers are "more like their predecessors who graduated decades earlier than they are like today's children". The result, according to Strommen and Lincoln (1992), is that there exists an estrangement of schools from both societies in general and from the children who attend these schools.

Appropriate Assessment in a Constructivist Setting

Appropriate assessment must be considered, if briefly. Jonassen (1991) states, "Perhaps the thorniest issue yet to be resolved regarding the implications of constructivism for learning is how to evaluate the learning that emerges from those environments. If constructivism is a valid perspective for delivering instruction, then it should also provide a valid set of criteria for evaluating the outcomes of that instruction. That is, the assumptions of constructivism should be applied to evaluation". Jonassen (1991) makes twelve points about appropriate assessment and constructivism:

1. Technology can and will force the issue of constructivism.
2. Assessment will have to be outcome based andstudent centered.

3. Assessment techniques must be developed which reflect instructional outcomes.
4. "Grades" must be contracted where grades are required.
5. There must be non-graded options and portfolio assessment.
6. There must be self and peer evaluation as well as teacher assessment.
7. Performance standards must be developed.
8. A grading system must be developed which provides meaningful feedback.
9. Technology will be used to facilitate communication with parents.
10. Students will be videotaped as they work as part of their portfolio.
11. The focus must be on originality rather than regurgitation; it is important to evaluate how the learner goes about constructing his or her own knowledge rather than the product.
12. Assessment is context dependent.

Because of the growing dependence and importance of technology in society, technology will become increasingly important in schools as problem-solving tools. In fact, the potential of new technologies has been generally unrealized by many educators. "Many educators, as well as members of the general community, are naive about the ramifications of technology implementation, and proceed without a clear understanding of both the role of technology in schools and what are reasonable expectations".

Personal Application of Constructivism and Technology

For many years I have worked with a small, but enlightened group of individuals who have spent many, many hours discussing ways to effect positive change in Montgomery County Public Schools. This dedicated group developed curriculum, provided staff development, acquired appropriate instructional materials, and lent support to teachers striving to make their classrooms more child centered.

As we set out to write this grant we began by looking at how children learn. We also looked at what Steve Hodas (1993) calls technology refusal. These two issues are the driving force for our proposal. Constructivism as a theory will be forced into play by emerging technologies because it is impossible for a teacher to use didactic methodology in a technology-rich classroom. Appropriate assessment will naturally follow as school reform incorporates technology into the curriculum.

The space science technology development program develops and makes available new space technologies needed to enable or enhance exploration, expand our knowledge of the universe, and ensure continued national scientific, technical, and economic leadership. It strives to improve reliability and mission safety, and to accelerate mission development. Since the early 1990's, the average space science mission development time has been reduced from over nine years to five years or less, partly by integration and early infusion of advanced technologies into missions. For missions planned in the years 2000 to 2004, we hope to further reduce development time to less than four years. Our technology program encompasses three key objectives. First, we strive to develop new and better technical approaches and capabilities. Where necessary, we then validate these capabilities in space so that they can be confidently applied to science flight projects. Finally, we use these improved and demonstrated capabilities in the science programs and ultimately transfer them to U.S. industry for public use.

To achieve the technology goals for meeting future space science missions' capability requirements formulated in science roadmaps, the Enterprise technology strategy is to:

- Focus technology development on science program requirements.

When near-term Enterprise mission concepts are defined sufficiently to begin detailed scoping of their instrumentation, systems, and infrastructure, performance requirements are derived. Technology development is focused on meeting the identified requirements ("mission pull" technologies). Basic technology research is focused on perceived longer range technology needs (they are characterized with less precision than near-term requirements). These longer range needs are generated from advanced mission concepts developed by mission study groups working with science advisory groups. Identified needs are then used to allocate support for maturation of more revolutionary technical approaches ("vision pull" technologies). The balance between "mission pull" and "vision pull" contributes to program agility and results in long-term continuing progress of the overall program notwithstanding short-term changes in circumstances.

- Fund technologies of high value to future Enterprise missions to the point that they are ready for ground or space demonstration.

A large number of technology concepts are given the benefit of exploratory research in the expectation that a fraction of these will emerge as promising. It is critical that these promising candidates for use in missions are identified and funded to the point where they are tested in a relevant environment on the ground, adopted by a flight project for further maturation, or are proposed as a candidate for space flight validation.

- Formulate and implement cost-effective space demonstrations of selected technologies on suitable carriers.

Project managers need assurance that adopted new technology will perform in the relevant space environment. In many cases this environment can be simulated on the ground in a satisfactory manner. Often this is not the case, however. It is necessary to identify those technologies that require space flight demonstration and perform these demonstrations.

Use these technologies in multiple missions to reduce costs and shorten mission development time across the program.

Since the early 1990's, the average science mission development time has been reduced from over nine years to five years. Although many factors can compress mission development time, infusion of validated new technology in the early mission phases can facilitate this. We compare requirements for future missions to look for common needs that can be met through a coordinated technology development effort. Where possible, we sequence missions so that later projects build on technology developed and successfully demonstrated for earlier ones.

- Promote partnerships with other agencies, industry, and academia to take advantage of external capabilities.

Technology infusion succeeds best through formal and informal interactions between mission developers, scientific principal investigators, and technology providers. In the early phases of development, detailed analyses and trade-off studies are conducted to determine technical feasibility and to establish technology priorities. Joint activities and partnerships with universities, industry, and other Government agencies can be particularly important in speeding identification and realization of the most attractive technology for specific needs. Technical capabilities can sometimes be efficiently acquired through cooperation with international partners. In conducting cooperative activities in technology, the Enterprise protects proprietary data and intellectual property.

- Maintain excellence by engaging the outside community in technology development and evaluation.

The Enterprise integrates industry, academia, and other Federal agencies' laboratories into the program. Industry, for example, develops valuable technology by using internal resources (IR&D and profit dollars), and through service to non-NASA customers. Similarly, the university community holds a vast and unique technological resource. University groups, receiving significant funding from other sources, are often leaders in essential technology areas. This approach of using a mix of dedicated peer-reviewed efforts at NASA Centers, other Government agencies, industry, and universities—ensures that the "best and brightest" are tapped for the required developments.

This approach dovetails with the practices of following parallel paths in early development followed by descoping and down-selecting. Independent merit review will be used to assure excellence in both internal NASA and external technology development efforts. These reviews will consider whether work is "best-in-class," contributes to specific, documented, and otherwise unaddressed Enterprise requirements, and is advancing significantly the state-of-the-art.

The role of science and technology in future design will be discussed from the perspective of someone who has lived all his life in the United States and whose scientific experience has spanned the years since the late 1930s. It is likely that the reader will find in my discussion characteristics that apply to many developed countries and developing ones. Inasmuch as scientific progress is highly dependent on financial support and, in modern times, on general societal support, it is appropriate to discuss the interaction of science and society. Using the United States as an example, some of the topics to be discussed are the views of public officials who influence the distribution of research funds, the response of funding agencies and the views of scientists. Finally, we shall look at the co-evolution of science and society and attempt to draw some conclusions concerning their related future and the implications for the future of technology.

Views of Public Officials

Public officials who are involved in setting or influencing science policy have expressed opinions that indicate that they intend to change the basis for supporting research and development. They speak in terms of a "paradigm shift" based on some new perception of the role of science in society. The word paradigm has several meanings, but in the way it is used here the words "pattern" or "model" may be good substitutes. In other words, the public officials wish to alter somewhat the pattern of funding for science. Their motivation is to orient research more toward programs that, for example, ensure a stronger economy and improvements in the environment. It is becoming increasingly apparent that those public officials, who control public funds, will be reluctant to fund research programs that they consider unrelated to national needs.

An example of priority-setting by public officials was the vote in the House of Representatives against further construction of the high energy accelerator known as the superconducting super collider. This shift in spending priorities implies that nuclear physics may receive less support in the future if it continues to be viewed as less related to the new national priorities than other scientific disciplines.

Views of Funding Agencies

The effect of the intention of federal officials to shift public research funds toward research programs that serve the national priorities has already affected the nature of the funding available at the funding agencies. For example, at the National Science Foundation, a small increase in funding for the chemistry division is directed toward so-called strategic research initiatives that involve, for example, advanced materials and processing, biotechnology, environmental chemistry and high-performance computing.

It is likely that this trend will continue. The Federal Coordinating Council on Science, Engineering and Technology identified the current national priority areas as high-performance computing, advanced materials, manufacturing research and education, biotechnology and global change. The expressed intention is to get more effort into those areas, but not to have them be entirely exclusive.

Views of Scientists

Many questions arose in the scientific community as a consequence of the use of words such as "new paradigm," "strategic areas", "priorities," and "national competitiveness" in statements concerning the future funding of science. The questions concerned many aspects of the support of science, such as, is the paradigm really new, who decides which areas are strategic and who sets the priorities, and are the important contributions of curiosity-driven basic research to be largely sacrificed. The indications so far are quite clear that the government expects to shift publicly funded research activity into the areas that are deemed strategic. Is this a new paradigm or merely a shift in emphasis?

Quite apparently there has been over the years heavy funding and much research in the strategic (priority) areas. There also has been in the United States, a major Industry-University cooperative research program conducted by the National Science Foundation. It celebrated its 20th year of operation in January, 1994. An account of this very successful and extensive program has been presented in the January 24, 1994 issue of *Chemical and Engineering News* published by the American Chemical Society. The motivation of this cooperative program is to develop and transfer industrially relevant technologies from the university into practice.

There are currently more than 50 active centers involving about 1,000 faculty members, about 1,000 graduate students and 78 universities. More than 700 organizations sponsor the centers, including government agencies, national laboratories and about 500 industrial firms. It is pointed out that the success rate is very high, namely only 6% of the centers have failed. Major investments have been made by sponsor organizations, based on center technologies. There are also many other industry-university collaborations that are not part of the National Science Foundation program.

Do we really have a "new paradigm" and, if so, what is it? Performing research in the interest of national needs is not new. Cooperating with industry is not new. Setting priorities is not new. What could be new? It is indicated that what is new is that by control of public funds curiosity driven research is to be curtailed to some unspecified degree in favor of research perceived to be in the national interest. This, I believe is the source of the apprehension among scientists. The major developments in science and technology generally derive from curiosity driven research and these developments have had over time great impact on the national interest, enriching the country with whole new industries and making contributions to the health, welfare, comfort and security of society. Is curtailing curiosity driven research in the national interest?

Impact of Curiosity Driven Basic Research

Many scientific groups have produced literature that describes, in terms of many examples, how curiosity driven research has led to important developments in the interest of society. The October, 1993 issue of *Physics Today* celebrated the one hundredth anniversary of the journal, *Physical Review.* A major part of this issue was devoted to the matter of basic research. An article by Robert K. Adair and Ernest M. Henley pointed out that "a century of fundamental physics research has appeared in the *Physical Review.* Such research is the seed corn of the technological harvest that sustains modern society." In an article on the laser, Nicolas Bloembergen points out that "the first paper reporting an operating laser was rejected by *Physical Review Letters* in 1960.

Now lasers are a huge and growing industry, but the pioneers' chief motivation was the physics." In an article on fiber optics, Alister M. Glass notes that "fundamental research in glass science, optics and quantum mechanics has matured into a technology that is now driving a communications revolution." In an article on superconductivity, Theodore H. Geballe states that "it took half a century to understand Kamerlingh Onnes' discovery, and another quarter-century to make it useful. Presumably we won't have to wait that long to make practical use of the new high-temperature superconductors." Other articles concerned nuclear magnetic resonance, semiconductors, nanostructures and medical cyclotrons, all subjects of great technological and medical importance that originated in basic physical research.

In a preface for a publication of the American Chemical Society, *Science and Serendipity*, the President of the ACS in 1992, Ernest L. Eliel, writes about "The Importance of Basic Research." He writes that "many people believe - having read about the life of Thomas Edison - that useful products are the result of targeted research, that is, of research specifically designed to produce a desired product. But the examples given in this booklet show that progress is often made in a different way. Like the princes of Serendip, researchers often find different, sometimes greater, riches than the ones they are seeking. For example, the tetrafluoroethylene cylinder that gave rise to Teflon was meant to be used in the preparation of new refrigerants.

And the anti-AIDS drug AZT was designed as a remedy for cancer." He goes on to say that "most research stories are of a different kind, however. The investigators were interested in some natural phenomenon, sometimes evident, sometimes conjectured, sometimes predicted by theory. Thus, Rosenberg's research on the potential effects of electric fields on cell division led to the discovery of an important cancer drug; Kendall's work on the hormones of the adrenal gland led to an anti-inflammatory substance; Carothers' work on giant molecules led to the invention of Nylon; Bloch and Purcell's fundamental work in the absorption of radio frequency by atomic nuclei in a magnetic field led to MRI. Development of gene splicing by Cohen and Boyer produced, among other products, better insulin. Haagen-Smit's work on air pollutants spawned the catalytic converter.

Reinitzer's discovery of liquid crystals is about to revolutionize computer and flat-panel television screens, and the discovery of the laser - initially a laboratory curiosity - is used in such diverse applications as the reattachment of a detached retina and the reading of barcodes in supermarkets. All of these discoveries are detailed in this booklet (*Science and Serendipity*). Ernest Eliel goes on to point, out that "the road from fundamental discovery to practical application is often quite long, ranging from about 10 years in the example of Nylon to some 80 years in the case of liquid crystals." He concludes that "if we stop doing fundamental research now, the 'well' that supplies the applications will eventually run dry. In other words, without continuing fundamental research, the opportunities for new technology are eventually going to shrink."

Some of the other topics in the brochure on *Science and Serendipity*, that were included to document further the importance of basic research, concerned several examples of the impact of chemistry on medicine. There are, in fact, countless such examples. The Federation of American Societies for Experimental Biology (FASEB) in their Newsletter of May, 1993 considered basic biomedical research and its benefits to society. I quote from the FASEB Public Affairs Bulletin of May, 1993. "There have been recent suggestions that tighter linkage between basic research and national goals should become a criterion for research support. Concerns also have been raised that science is being practiced for its own sake, and that it would be better for the nation if research were oriented more toward specific industrial applications."

They go on to point out that "the available evidence, however, clearly indicates that the desired linkage already exists. Indeed, a majority of scientists are intimately involved in the study and treatment of common human diseases and collaborate closely with clinical scientists. Industries involved in biomedical development have been remarkably efficient in commercial application of treatment modalities based on discoveries resulting from fundamental research funded primarily by the federal government. "A critical factor in sustaining the competitive position of biomedical-based industries is for basic research to continue to provide a stream of ideas and discoveries that can be translated into new products. It is essential to provide adequate federal support for a broad base of fundamental research, rather than shifting to a major emphasis on directed research, because the paths to success are unpredictable and subject to rapid change.

"History has repeatedly demonstrated that it is not possible to predict which efforts in fundamental research will lead to critical insights about how to prevent and treat disease; it is therefore essential to support a sufficient number of meritorious projects in basic research so that opportunities do not go unrealized. Although its primary aim is to fill the gaps in our understanding of how life processes work, basic research has borne enormous fruit in terms of its practical applications. We recognize that during a time when resources are constrained, it may be tempting to direct funding to projects that appear likely to provide early practical returns, but we emphasize that support for a wide-ranging portfolio of untargeted research has proven to be the better investment.

This provides the broader base of knowledge from which all new medical applications arise. Decisions regarding what research to fund must be based on informed judgments about which projects represent the most meritorious ideas." FASEB continues with a discussion of economic benefits and a number of examples of basic research-driven medical breakthroughs. "Society reaps substantial benefit from basic research. Technologies derived from basic research have saved millions of lives and billions of dollars in health care costs. According to an estimate by the National Institutes of Health on the economic benefits of 26 recent advances in the diagnosis and treatment of disease, some $6 billion in medical costs are saved annually by those innovations alone.

The significance of these basic research-derived developments, however, transcends the lowering of medical costs: the lives of children as well as adults are saved, and our citizens are spared prolonged illness or permanent disability. Fuller, more productive lives impact positively on the nation's economic and social progress." FASEB continues with thirteen examples of contributions by basic research to the diagnosis and treatment of numerous diseases, most of them very serious. Also noted in this Public Affairs Bulletin is that "our ability to know in advance all that is relevant is very poor" and that, in suggesting new ideas for the management of funding for science, never considered were "the serious consequences of harming the system."

Technology

Up to this point, we have been concerned with basic science and its support by government funds in a modern society. Although there is also some support by private institutions established for that purpose and also some industrial investment in generally product-oriented basic research, the greatest amount of support by far comes from public funds. One of the ways that the public is repaid for their support is through the technology that fundamental research generates. I suspect that the economic return from technology alone more than compensates for the monies expended for the entire basic research effort. I have no estimate, however, of whether my suspicion is true or not. It should be noted that the public gains much more than the economic value of technology. It gains culture, comfort, convenience, security, recreation, health and the extension of life. What monetary value can be put on the triumphs of health over debilitating or fatal disease? The monetary value has to be higher than the purely economic savings that were noted above in the 26 examples referred to in the FASEB Bulletin.

The word "technology" means industrial science and is usually associated with major activities such as manufacturing, transportation and communication. Technology has been, in fact, closely associated with the evolution of man starting with tools, clothing, fire, shelter and various other basic survival items. The co-evolution persists and, since basic science is now very much a part of developing technologies, the term co-evolution of science and society which is used at times very much implies the co-evolution of both basic science and industrial science with society. Advances in technology are generally accompanied by social changes as a consequence of changing economies and ways of carrying out life's various activities.

An important question arises concerning how basic scientific discoveries eventually lead to new technologies and what that may mean to the rational support of basic research and the future of science and technology in the developed and developing world. There are great uncertainties in the process that starts with basic research and ends with an economically successful technology. The successful discovery of a new development in research that appears to have technological significance does not ensure the economic success of technologies that may be based on it. Nathan Rosenberg of Stanford University, in a speech, "Uncertainty and Technological Change", before the National Academy of Sciences (April, 1994), pointed out that there are great uncertainties regarding economic success even in research that is generally directed toward a specific technological goal.

He notes that uncertainties derive from many sources, for example, failure to appreciate the extent to which a market may expand from future improvement of the technology, the fact that technologies arise with characteristics that are not immediately appreciated, and failure to comprehend the significance of improvements in complementary inventions, that is inventions that enhance the potential of the original technology. Rosenberg also points out that many new technological regimes take many years before they replace an established technology and that technological revolutions are never completed overnight. They require a long gestation period. Initially it is very difficult to conceptualize the nature of entirely new systems that develop by evolving over time. Rosenberg goes on to note that major or "breakthrough" innovations induce other innovations and their "ultimate impact depends on identifying certain specific categories of human needs and catering to them in novel or more cost effective ways. New technologies need to pass an economic test, not just a technological one."

What does this mean with regard to government managed research? I quote from Rosenberg's speech. "I become distinctly nervous when I hear it urged upon the research community that it should unfurl the flag of 'relevance' to social and economic needs. The burden of much of what I said is that we frequently simply do not know what new findings may turn out to be relevant, or to what particular realm of human activity that relevance may eventually apply. Indeed, I have been staking the broad claim that a pervasive uncertainty characterizes, not just basic research, where it is generally acknowledged, but the realm of product design and new product development as well - i.e., the D of R&D.

Consequently, early pre-commitment to any specific, large-scale technology project, as opposed to a more limited, sequential decision-making approach, is likely to be hazardous - i.e., unnecessarily costly. Evidence for this assertion abounds in such fields as weapons procurement, the space program, research on the development of an artificial heart, and synthetic fuels. "The pervasiveness of uncertainty suggests that the government should ordinarily resist the temptation to play the role of a champion of any one technological alternative, such as nuclear power, or any narrowly concentrated focus of research support, such as the War on Cancer.

Rather, it would seem to make a great deal of sense to manage a deliberately diversified research portfolio, a portfolio that will illuminate a range of alternatives in the event of a reordering of social or economic priorities. My criticism of the federal government's postwar energy policy is not that it made a major commitment to nuclear power that subsequently turned out to be problem-ridden. Rather, the criticism is aimed at the single-mindedness of the focus on nuclear power that led to a comparative neglect of many other alternatives, including not only alternative energy sources but improvements in the efficiency of energy utilization."

To these words, I add those (noted by FASEB) of Bruce Ferguson, Executive Vice President of Orbital Sciences Corporation, and a space technology firm. Ferguson said, "The federal government should focus its research and development spending on those areas for which the benefits are diffuse and likely to be realized over many years, rather than areas for which benefits are concentrated on particular products or firms over a few years. These areas are not well covered by corporate investment, yet are vital to the long-term economic strength of the country."

Some reactions to "strategic" research are recounted in an article in *Nature* of February 10, 1994 from which I quote some passages. The concept of strategic research "is not an unfamiliar cry, witness last year's debate in Britain about harnessing of research to 'wealth creation.' Nor, of course, is the objective in any way disreputable; what scientist would not be cheered to know that his or her research won practical benefits for the wider world as well as a modicum of understanding? The difficulties are those of telling in advance which particular pieces of research will lead to 'new technologies' and then to 'jobs'. "The recent past is littered with examples of adventurous goal-directed programmes of research and development which have failed for intrinsic reasons or which, alternatively, have been technically successful, but unusable for economic or other reasons."

The article goes on to say that the affection for strategic research in the United States may prove short-lived. "In Britain, much the same seems to be happening. Having pinned its reorganization of research on the doctrine of science for wealth-creation, the government appears now to be more conscious of the problems it has undertaken to solve. Indeed, the Prime Minister, John Major, seemed to be suggesting in a speech last week that the British part of the research enterprise deserves respect of the kind accorded to other social institutions at the heart of his 'back to basics' rhetoric. After more than a decade of needless damage-doing, that would be only prudent."

As a final remark, the article ends with the statement: "On the grander questions, on both sides of the Atlantic, it seems likely that the first flush of enthusiasm for turning research into prosperity will be abated by the reality of the difficulties of doing so. When governments discover in the course of seeking radical reorganization that the best they can do with their parts of the research enterprise is to cherish them, the lessons are likely to be remembered. If the outcome in the research community is a more vivid awareness of how much the world at large looks to research for its improvement, so much the better."

The Future of Science, Technology and Society

In discussing the future of science (including industrial science) and society, it is valuable to recount some of the important points that emerged from the previous discussion. As a consequence of recognizing the economic benefits that derive from the development of novel, successful technologies, governments have been attempting to direct research, supported with public funds, toward subjects that are perceived as national priorities. This contrasts with broadbased "curiosity" oriented basic research.

The views of scientists, a distinguished economist, some industrial leaders and an editorial comment in a distinguished science journal provide very strong indications that governmental management of goal-oriented research is replete with uncertainties and pitfalls and, although well-motivated, may cause serious damage to the scientific culture. This, of course, would defeat the original purpose, since the co-evolution of science and society is a very-well documented and irrefutable phenomenon. Strong arguments are presented in this chapter by individuals and groups that support the current system of governmental funding of a very broad range of scientific efforts as probably being as close to optimal with regard to national priorities as is possible.

No one can predict with any certainty what the most successful inventions and technologies will be in the future. The economic return on federally supported funding was the subject of a report by the Council of Economic Advisors to President Clinton. This report was released in November 1995. It documents high returns to the economy and the importance of governmental involvement. By any measure, basic scientific research has made monumental contributions to technology and national priorities. The bond between basic research and the development of both novel and current technologies has been and is well in place.

There is no question that science and society will continue to co-evolve. The nature of this evolution will certainly be affected by the extent to which governments set funding priorities. Societies whose governments recognize the dependence of the development of successful novel technologies on broadly supported basic research are more likely to be healthier and economically prosperous in the future than those that does not. Because of the unpredictability of the details of the new science and technology that will evolve, the details of social evolution are also unpredictable.

4

Technology
The Opiate of the Intellectuals

If religion was formerly the opiate of the masses, then surely technology is the opiate of the educated public today, or at least of its favorite authors. No other single subject is so universally invested with high hopes for the improvement of mankind generally and of Americans in particular. The content of these millennial hopes varies somewhat from author to author, though with considerable overlap. A representative but by no means complete list of these promises and their prophets would include: an end to poverty and the inauguration of permanent prosperity, universal equality of opportunity, a radical increase in individual freedom, the replacement of work by leisure for most of mankind, fresh water for desert dwellers, permanent but harmless social revolution, the final comeuppance of Mao Tse-tung and all his ilk (same prophet), the triumph of wisdom over power, and, lest we forget, the end of ideology.

These hopes for mankind's, or technology's, future, however, are not unalloyed. Technology's defenders, being otherwise reasonable men, are also aware that the world population explosion and the nuclear missiles race are also the fruit of the enormous advances made in technology during the past half century or so. But here too a cursory reading of their literature would reveal widespread though qualified optimism that these scourges too will fall before technologies might. Thus population (and genetic) control and permanent peace are sometimes added to the already imposing roster of technology's promises. What are we to make of such extravagant optimism?

Several months ago Harvard University's Program on Technology and Society, "...an inquiry in depth into the effects of technological change on the economy, on public policies, and on the character of society, as well as into the reciprocal effects of social progress on the nature, dimension, and directions of scientific and technological development," issued its Fourth Annual Report to the accompaniment of full front-page coverage in *The New York Times* (January 18). Within the brief (fewer than 100) pages of that report and most clearly in the concluding essay by the Program's Director, Emmanuel G. Mesthene, one can discern some of the important threads of belief which bind together much current writing on the social implications of technology.

Mesthene's essay is worth extended analysis because these beliefs are of interest in themselves and, of greater importance, because they form the basis not of a new but of a newly aggressive rightwing ideology in this country, an ideology whose growing importance was accurately measured by the magnitude of the *Times*'s news report. At the very beginning of Mesthene's essay, which attempts to characterize the relationships between technological and social change, the author is careful to dissociate himself from what he believes are several extreme views of those relationships.

For example, technology is neither the relatively "unalloyed blessing" which, he claims, Marx, Comte, and the Air Force hold it to be, nor an unmitigated curse, a view he attributes to "many of our youth." (This is but the first of several reproofs Mesthene casts in the direction of youth.) Having denounced straw men to the right and left of him he is free to pursue that middle or moderate course favored by virtually all political writers of the day. This middle course consists of an extremely abstract and—politically speaking—sanitary view of technology and technological progress.

For Mesthene, it is characteristic of technology that it:

> ...creates new possibilities for human choice and action but leaves their disposition uncertain. What its effects will be and what ends it will serve are not inherent in the technology, but depend on what man will do with technology. Technology thus makes possible a future of open-ended options....

This essentially optimistic view of the matter rests on the notion that technology is merely "...the organization of knowledge for practical purposes..." and therefore cannot be purely boon or wholly burden. The matter is somewhat more complex:

> New technology creates new opportunities for men and societies and it also generates new problems for them. It has both positive and negative effects, and it usually has the two *at the same time and in virtue of each other.*

This dual effect he illustrates with an example drawn from the field of medicine. Recent advances there have created two new opportunities: (1) they have made possible treatment and cures that were never possible before, and (2) they provide a necessary condition for the delivery of adequate medical care to the population at large as a matter of right rather than privilege.

Because of the first, however, the medical profession has become increasingly differentiated and specialized and is tending to concentrate its best efforts in a few major, urban centers of medical excellence. Mesthene clearly intends but does not state the corollary to this point, namely that the availability of adequate medical care is declining elsewhere. Moreover, because of the second point, there have been...big increases in demand for medical services, partly because a healthy population has important economic advantages in a highly industrialized society. This increased demand accelerates the process of differentiation and multiplies the levels of paramedical personnel between the physician at the top and the patient at the bottom of the hospital pyramid.

Similarly, Mesthene points out those marvelous improvements in auto and air transportation have aggravated social and other problems in the inner city. Furthermore, Mass communications technology has also made rapid strides since World War II, with great benefit to education, journalism, commerce and sheer convenience. It has also been accompanied by an aggravation of social unrest, however, and may help to explain the singular rebelliousness of a youth that can find out what the world is like from television before home and school have had the time to instill some ethical sense of what it could or should be like.

Mesthene believes there are two distinct problems in technology's relation to society, a positive one of taking full advantage of the opportunities it offers and the negative one of avoiding unfortunate consequences which flows from the exploitation of those opportunities. Positive opportunities may be missed because the costs of technological development outweigh likely benefits. Mesthene seems convinced, however, that a more important case is that in which...technology lies fallow because existing social structures are inadequate to exploit the opportunities it offers. This is revealed clearly in the examination of institutional failure in the ghetto carried on by the Program. At point after point, ...analyses confirm...that existing institutions and traditional approaches are by and large incapable of coming to grips with the new problems of our cities—many of them caused by technological change...—and unable to realize the possibilities for resolving them that are also inherent in technology.

Vested economic and political interests serve to obstruct adequate provision of low-cost housing. Community institutions wither for want of interest and participation by residents. City agencies are unable to marshall the skills and take the systematic approach needed to deal with new and intensified problems of education, crime control, and public welfare. Business corporations, finally, which are organized around the expectation of private profit, are insufficiently motivated to bring new technology and management know-how to bear on urban projects where the benefits will be largely social.

His diagnosis of these problems is generous in the extreme:

All these factors combine to dilute what may be otherwise a genuine desire to apply our best knowledge and adequate resources to the resolution of urban tensions and the eradication of poverty in the nation.

Moreover, because government and the media "...are not yet equipped for the massive task of public education that is needed..." if we are to exploit technology more fully, many technological opportunities are lost because of the lack of public support. This too is a problem primarily of "institutional innovation."

Mesthene believes that institutional innovation is no less important in combatting the negative effects of technology. Individuals or individual firms who decide to develop new technologies normally do not take "adequate account" of their likely social benefits or costs. His critique is anti-capitalist in spirit, but lacks bite, for he goes on to add that ... most of the negative] consequences of technology that are causing concern at the present time—pollution of the environment, potential damage to the ecology of the planet, occupational and social dislocations, threats to the privacy and political significance of the individual, social and psychological malaise—are *negative externalities of this kind.* They are with us in large measure because it has not been anybody's explicit business to foresee and anticipate them.

Mesthene's abstract analysis and its equally abstract diagnosis in favor of "institutional innovation" place him in a curious and, for us, instructive position. If existing social structures are inadequate to exploit technology's full potential, or if, on the other hand, so-called "negative externalities" assail us because it is nobody's business to foresee and anticipate them; doesn't this say that we should apply technology to this problem too? That is, we ought to apply and organize the appropriate *organizational* knowledge for the practical purpose of solving the problems of institutional inadequacy and "negative externalities." Hence, in principle, Mesthene is in the position of arguing that the cure for technology's problems, whether positive or negative, is still more technology. This is the first theme of the technological school of writers and its ultimate First Principle.

Technology, in their view, is a self-correcting system. Temporary oversight or "negative externalities" will and should be corrected by technological means. Attempts to restrict the free play of technological innovation are, in the nature of the case, self-defeating. Technological innovation exhibits a distinct tendency to work for the general welfare in the long run. *Laissez innover!*

I have so far deliberately refrained from going into any greater detail than does Mesthene on the empirical character of contemporary technology for it is important to bring out the force of the principle of *laissez innover* in its full generality. Many writers on technology appear to deny in their definition of the subject—organized knowledge for practical purposes—that contemporary technology exhibits distinct trends which can be identified or projected. Others, like Mesthene, appear to accept these trends, but then blunt the conclusion by attributing to technology so much flexibility and "scientific" purity that it becomes an abstraction infinitely malleable in behalf of good, pacific, just, and egalitarian purposes. Thus the analogy to the *laissez-faire* principle of another time is quite justified.

Just as the market or the free play of competition provided in theory the optimum long-run solution for virtually every aspect of virtually every social and economic problem, so too does the free play of technology, according to its writers. Only if technology or innovation (or some other synonym) is allowed the freest possible reign, they believe, will the maximum social good be realized. What reasons do they give to believe that the principle of *laissez innover* will normally function for the benefit of mankind rather than, say, merely for the benefit of the immediate practitioners of technology, their managerial cronies, and for the profits accruing to their corporations? As Mesthene and other writers of his school are aware, this is a very real problem, for they all believe that the normal tendency of technology is, and ought to be, the increasing concentration of decision-making power in the hands of larger and larger scientific-technical bureaucracies. *In principle*, their solution is relatively simple, though not often explicitly stated.

Their argument goes as follows: the men and women who are elevated by technology into commanding positions within various decision-making bureaucracies exhibit no generalized drive for power such as characterized, say, the landed gentry of pre-industrial Europe or the capitalist entrepreneur of the last century. For their social and institutional position and its supporting culture as well are defined solely by the fact that these men are problem solvers. (Organised knowledge for practical purposes again.) That is, they gain advantage and reward only to the extent that they can bring specific technical knowledge to bear on the solution of specific technical problems. Any more general drive for power would undercut the bases of their usefulness and legitimacy.

Moreover their specific training and professional commitment to solving technical problems creates a bias against ideologies in general which inhibits any attempts to formulate a justifying ideology for the group.

Consequently, they do not constitute a class and have no general interests antagonistic to those of their problem-beset clients. We may refer to all of this as the disinterested character of the scientific-technical decision-maker, or, more briefly and cynically, as the principle of the Altruistic Bureaucrat. As if not satisfied by the force of this (unstated) principle, Mesthene like many of his school fellows spends many pages commenting around the belief that the concentration of power at the top of technology's organizations is a problem, but that like other problems technology should be able to solve it successfully through institutional innovation.

You may trust in it; the principle of *laissez innover* knows no logical or other hurdle. This combination of guileless optimism with scientific tough mindedness might seem to be no more than an eccentric delusion was the American technology it supports not moving in directions that are strongly anti-democratic. To show this we must examine more closely Mesthene's seemingly innocuous distinction between technology's positive opportunities and its "negative externalities." In order to do this I will make use of an example drawn from the very frontier of American technology, the Vietnam War.

At least two fundamentally different bombing programs are now being carried out in South Vietnam. There are fairly conventional attacks against targets which consist of identified enemy troops, fortifications, medical centers, vessels, and so forth. The other program is quite different and, at least since March, 1968, infinitely more important. With some oversimplification it can be described as follows:

Intelligence data is gathered from all kinds of sources, of all degrees of reliability, on all manner of subjects, and fed into a computer complex located, I believe, at Bien Hoa. From this data and using mathematical models developed for the purpose, the computer then assigns probabilities to a range of potential targets, probabilities which represent the likelihood that the latter contain enemy forces or supplies.

These potential targets might include: a canal-river crossing known to be used occasionally by the NLF; a section of trail which would have to be used to attack such and such an American base, now overdue for attack; a square mile of plain rumored to contain enemy troops; a mountainside from which camp fire smoke was seen rising. Again using models developed for the purpose, the computer divides pre-programmed levels of bombardment among those potential targets which have the highest probability of containing actual targets. Following the raids, data provided by further reconnaissance is fed into the computer and conclusions are drawn (usually optimistic ones) on the effectiveness of the raids. This estimate of effectiveness then becomes part of the data governing current and future operations, and so on.

Two features must be noted regarding this program, features which are superficially hinted at but fundamentally obscured by Mesthene's distinction between the abstractions of positive opportunity and "negative externality." First, when considered from the standpoint of its planners, the bombing program is extraordinarily rational, for it creates previously unavailable "opportunities" to pursue their goals in Vietnam. It would make no sense to bomb South Vietnam simply at random, and no serious person or Air Force General would care to mount the effort to do so. So the system employed in Vietnam significantly reduces, though it does not eliminate, that randomness. That canal-river crossing which is bombed at least once every eleven days or so is a very poor target compared to an NLF battalion observed in a village. But it is an infinitely more promising target than would be selected by throwing a dart at a grid map of South Vietnam. In addition to bombing the battalion, why not bomb the canal crossing to the frequency and extent that it *might* be used by enemy troops?

Even when we take into account the crudity of the mathematical models and the consequent slapstick way in which poor information is evaluated; it is a "good" program.

No single raid will definitely kill an enemy soldier but a whole series of them increases the "opportunity" to kill a calculable number of them (as well, of course, as a calculable but not calculated number of non-soldiers). This is the most rational bombing system to follow if American lives are very expensive and American weapons and Vietnamese lives very cheap. Which, of course, is the case?

Secondly, however, considered from the standpoint of goals and values not programmed in by its designers, the bombing program is incredibly irrational. In Mesthene's terms, these "negative externalities" would include, in the present case, the lives and well-being of various Vietnamese as well as the feelings and opinions of some less important Americans. Significantly, this exclusion of the interests of people not among the managerial class is based quite as much on the so-called "technical" means being employed as on the political goals of the system. In the particular case of the Vietnamese bombing system, the political goals of the bombing system clearly exclude the interests of certain Vietnamese.

After all, the victims of the bombardment are communists or their supporters, they are our enemies, and they resist US intervention. In short, their interests are fully antagonistic to the goals of the program and simply must be excluded from consideration. The technical reasons for this exclusion require explanation, being less familiar and more important, especially in the light of Mesthene's belief in the malleability of technological systems. Advanced technological systems such as those employed in the bombardment of South Vietnam make use not only of extremely complex and expensive equipment but, quite as important, of large numbers of relatively scarce and expensive-to-train technicians. They have immense capital costs; a thousand aircraft of a very advanced type, literally hundreds of thousands of spare parts, enormous stocks of rockets, bombs, shells and bullets, in addition to tens of thousands of technical specialists; pilots, bombardiers, navigators, radar operators, computer programmers, accountants, engineers, electronic and mechanical technicians, to name only a few. In short, they are "capital intensive."

Moreover, the coordination of this immense mass of esoteric equipment and its operators in the most effective possible way depends upon an extremely highly developed technique both in the employment of each piece of equipment by a specific team of operators and in the management of the program itself. Of course, all large organizations standardize their operating procedures, but it is peculiar to advanced technological systems that their operating procedures embody a very high degree of information drawn from the physical sciences, while their managerial procedures are equally dependent on information drawn from the social sciences. We may describe this situation by saying that advanced technological systems are both "technique intensive" and "management intensive."

It should be clear, moreover, even to the most casual observer that such intensive use of capital, technique, and management spills over into almost every area touched by the technological system in question. An attack program delivering 330,000 tons of munitions more or less selectively to several thousand different targets monthly would be an anomaly if forced to rely on sporadic intelligence data, erratic maintenance systems, or a fluctuating and unpredictable supply of heavy bombs, rockets, jet fuel, and napalm tanks. Thus it is precisely because the bombing program requires an intensive use of capital, technique, and management that the same properties are normally transferred to the intelligence, maintenance, supply, coordination and training systems which support it. Accordingly, each of these supporting systems is subject to sharp pressures to improve and rationalize the performance of its machines and men, the reliability of its techniques, and the efficiency and sensitivity of the management controls under which it operates. Within integrated technical systems, higher levels of technology drive out lower, and the normal tendency is to integrate systems.

From this perverse Gresham's Law of Technology follow some of the main social and organizational characteristics of contemporary technological systems: the radical increase in the scale and complexity of operations that they demand and encourage; the rapid and widespread diffusion of technology to new areas; the great diversity of activities which can be directed by central management; an increase in the ambition of management's goals; and, as a corollary, especially to the last, growing resistance to the influence of so-called "negative externalities." Complex technological systems are extraordinarily resistant to intervention by persons or problems operating outside or below their managing groups, and this is so regardless of the "politics" of a given situation. Technology creates its own politics. The point of such advanced systems is to minimize the incidence of personal or social behavior which is erratic or otherwise not easily classified, of tools and equipment with poor performance, of improvisory techniques, and of unresponsiveness to central management.

For example, enlisted men who are "unrealistically soft" on the subject of civilian casualties and farmers in contested districts pose a mortal threat to the integral character of systems like that used in Vietnam. In the case of the soldier this means he must be kept under tight military discipline. In the case of the farmer, he must be easily placed in one of two categories; collaborator or enemy. This is done by assigning a probability to him, his hamlet, his village, or his district, and by incorporating that probability into the targeting plans of the bombing system. Then the enlisted man may be controlled by training and indoctrination as well as by highly developed techniques of command and coercion, and the farmers may be bombed according to the most advanced statistical models. In both cases the system's authority over its farmer subjects or enlisted men is a technical one.

The technical means which make that system rational and efficient in its aggregate terms, i.e., as viewed from the top, themselves tend by design to filter out the "non-rational" or "non-efficient" elements of its components and subjects, i.e., those rising from the bottom. To define technology so abstractly that it obscures these observable characteristics of contemporary technology—as Mesthene and his school have done—makes no sense. It makes even less sense to claim some magical malleability for something as undefined as "institutional innovation."

Technology, in its concrete, empirical meaning, refers fundamentally to systems of rationalized control over large groups of men, events, and machines by small groups of technically skilled men operating through organizational hierarchy. The latent "opportunities" provided by that control and its ability to filter out discordant "negative externalities" are, of course, best illustrated by extreme cases. Hence the most instructive and accurate example should be of a technology able to suppress the humanity of its rank-and-file and to commit genocide as a by-product of its rationality. The Vietnam bombing program fits technology to a "T."

It would certainly be difficult to attempt to translate in any simple and direct way the social and organizational properties of highly developed technological systems from the battle-fields of Vietnam to the different cultural and institutional setting of the US. Yet before we conclude that any such attempt would be futile or even absurd, we might consider the following story. In early 1967 I stayed for several days with one of the infantry companies of the US Fourth Division whose parent battalion was then based at Dau Tieng. From the camp at Dau Tieng the well-known Black Lady Mountain, sacred to the Cao Dai religious sect, was easily visible and in fact dominated the surrounding plain and the camp itself. One afternoon when I began to explain the religious significance of the mountain to some GI friends, they interrupted my somewhat academic discourse to tell me a tale beside which even the strange beliefs of the Cao Dai sect appeared prosaic.

According to GI reports which the soldiers had heard and believed, the Viet Cong had long ago hollowed out most of the mountain in order to install very big cannon there. The size of the cannon was left somewhat vague, but clearly the GI's imagined that it was in the battleship class. In any event, these huge cannon had formerly taken a heavy toll of American aircraft and had been made impervious to American counterattacks by the presence of two—"huge, fucking"—sliding steel doors, behind which it retreated whenever the Americans attacked. Had they seen these battleship cannon, and did it ever fire on the camp, which was easily within its range? No, they answered, for a brave flyer, recognizing the effectiveness of the cannon against his fellow pilots, had deliberately crashed his jet into those doors one day, jamming them, and permitting the Americans to move into the area unhindered.

Only later, after talking to many soldiers and hearing many other wild stories from them as well, did I realize what the explanation for this was. Unlike officers and civilian correspondents who are almost daily given detailed briefings on a unit's situation capabilities and objectives, GI's are told virtually nothing of this sort by the Army. They are simply told what to do, where, and how, and it is a rare officer, in my experience anyway, who thinks they should be told any more than this. Officers don't think soldiers are stupid; they simply assume it, and act accordingly. For the individual soldier's personal life doesn't make too much difference; he still has to deal with the facts of personal feelings, his own well-being, and that of his family.

But for the soldier's group life this makes a great deal of difference. In their group life, soldiers are cut off from sources of information about the situation of the group and are placed in a position where their social behavior is governed largely by the principle of blind obedience. Under such circumstances, reality becomes elusive. Because the soldiers are not permitted to deal with facts in their own ways, facts cease to discipline their opinions. Fantasy and wild tales are the natural outcome. In fact, it is probably a mark of the GI's intelligence to fantasize, for it means that he has not permitted his intellectual capacity to atrophy. The intelligence of the individual is thus expressed in the irrationality of the group.

It is this process which we may observe when we look to the social effect of modern technological systems in America itself. Here the process is not so simple and clear as in Vietnam, for it involves not simply the relations of today's soldiers to their officers and to the Army but the historical development of analogous relations between the lower and upper orders of our society. Moreover, these relations are broadly cultural rather than narrowly social in nature. It is to a brief review of this complex subject that I now wish to turn.

Among the conventional explanations for the rise and spread of the democratic ethos in Europe and North America in the seventeenth, eighteenth, and nineteenth centuries, the destruction of the gap in political culture between the mass of the population and that of the ruling classes is extremely important. There are several sides to this explanation. For example, it is often argued that the invention of the printing press and the spread of Protestant Christianity encouraged a significant growth in popular literacy. In its earliest phases this literacy was largely expended on reading the Old and New Testament, but it quickly broadened to include other religious works such as Bunyan's *Pilgrim's Progress,* and after that to such secular classics as *Gulliver's Travels.*

The dating of these developments is, in the nature of the case, somewhat imprecise. But certainly by the middle of the eighteenth century, at least in Britain and North

America, the literacy of the population was sufficient to support a variety of newspapers and periodicals not only in the larger cities but in the smaller provincial towns as well. The decline of Latin as the first language of politics and religion paralleled this development, of course. Thus, even before the advent of Tom Paine, Babeuf, and other popular tribunes, literacy and the information it carried were widely and securely spread throughout the population and the demystification of both the religious and the political privileges of the ruling classes was well developed. Common townsmen had closed at least one of the cultural gaps between themselves and the aristocracy of the large cities.

Similarly, it is often argued that with the expansion and improvement of road and postal systems, the spread of new tools and techniques, the growth in the number and variety of merchants, the consequent invigoration of town life, and other numerous and familiar related developments, the social experiences of large numbers of people became richer, more varied, and similar in fact to those of the ruling classes. This last, the growth in similarity of the social experiences of the upper and lower classes, is especially important. Social skills and experiences which underlay the monopoly of the upper classes over the processes of law and government were spreading to important segments of the lower orders of society.

For carrying on trade, managing a commercial—not a subsistence—farm, participating in a vestry or workingmen's guild, or working in an up-to-date manufactory or business, unlike the relatively narrow existence of the medieval serf or artisan, were experiences which contributed to what I would call the social rationality of the lower orders. Activities which demand frequent intercourse with strangers, accurate calculation of near means and distant ends, and a willingness to devise collective ways of resolving novel and unexpected problems demand and reward a more discriminating attention to the realities and deficiencies of social life, and provide thereby a rich variety of social experiences analogous to those of the governing classes.

As a result not only were the processes of law and government, formerly treated with semi-religious veneration, becoming demystified but, equally important, a population was being fitted out with sufficient skills and interests to contest their control. Still another gap between the political cultures of the upper and lower ends of the social spectrum was being closed. The same period also witnesses a growth in the organized means of popular expression. In Britain, these would include the laboring people's organizations whose development is so ably described in Edward Thompson's *The Making of the English Working Class*. In America, the increase in the organized power of the populace was expressed not only in the growing conflict between the colonies and the Crown but more sharply and fundamentally in the continuous antagonism between the coastal areas and the backwoods, expressed, for example, in Shay's rebellion in western Massachusetts in 1786.

Clearly these organizational developments were related to the two foregoing as both cause and effect. For the English workingmen's movement and the claims to local self-government in America spurred, and were spurred by, the growth in individual literacy and in social rationality among the lower classes. They were in fact its organizational expression. These same developments were also reflected in the spread of egalitarian and republican doctrines such as those of Richard Price and Thomas Paine, which pointed up the arbitrary character of what had heretofore been considered the rights of the higher orders of society, and thus provided the popular ideological base which helped to define and legitimate lower-class demands.

This description by no means does justice to the richness and variety of the historical process underlying the rise and spread of what has come to be called the democratic ethos. But it does, I hope, isolate some of the important structural elements and, moreover, it enables us to illuminate some important ways in which the new technology, celebrated by Mesthene and his associates for its potential contributions to democracy, contributes instead to the erosion of that same democratic ethos. For if, in an earlier time, the gap between the political cultures of the higher and lower orders of society was being widely attacked and closed, this no longer appears to be the case. On the contrary, I am persuaded that the direction has been reversed and that we now observe evidence of a growing separation between ruling and lower class culture in America, a separation which is particularly enhanced by the rapid growth of technology and the spreading influence of its *laissez innover* ideologues.

Certainly, there has been a decline in popular literacy, that is to say, in those aspects of literacy which bear on an understanding of the political and social character of the new technology. Not one person in a hundred is even aware of, much less understands, the nature of technologically highly advanced systems such as are used in the Vietnam bombing program. People's ignorance in these things is revealed in their language. No clearer illustration of this ignorance is needed than the growing and already enormous difference between the speech of organizational and technical specialists and that of the man in the street, including many of the educated ones. To the extent that technical forms of speech within which the major business of American society is carried on are not understood or are poorly understood, there is a decline in one of the essentials of democracy.

This is not to say that the peculiar jargon which characterizes the speech of, say, aerospace technicians, crisis managers, or economic mandarins are intrinsically superior to the vocabulary of ordinary conversation, though sometimes this is indeed the case. What is important about technical language is that the words, being alien to ordinary speech, hide their meaning from ordinary speakers; terms like foreign aid or technical assistance have a good sound in ordinary speech; only the initiate recognizes them as synonyms for the old-fashioned, nasty word, imperialism. Such instances can be corrected but when almost all of the public's business is carried on in specialized jargon correction makes little difference.

Like Latin in the past, the new language of social and technical organization is divorced from the general population, which continues to speak in the vulgar tongue of, say, *The New Republic*, the *Saturday Review of Literature*, or *The Reader's Digest*. Secondly, the social organization of the new technology, by systematically denying to the general population experiences which are analogous to those of its higher management, contributes very heavily to the growth of social irrationality in our society. For example, modern technological organization defines the roles and values of its members, not vice versa. An engineer or a sociologist is one who does all those things but only those things called for by the "table of organization" and the "job description" used by his employer. Professionals who seek self-realization through creative and autonomous behavior without regard to the defined goals, needs, and channels of their respective departments have no more places in a large corporation or government agency than squeamish soldiers in the Army.

Naturally some tolerance would normally be extended to very gifted or personable individuals. This is especially true in universities. But for the common garden variety employee (or junior faculty member) company sanctions on job behavior, style of work, and related matters must have the force of law. However, those at the top of technology's more advanced organizations hardly suffer the same experience. For reasons which are clearly related to the principle of the Altruistic Bureaucracy the psychology of an individual's fulfillment through work has been incorporated into management ideology. As the pages of *Fortune, Time,* or *Business Week* or the memoirs of out-of-office Kennedyites serve to show, the higher levels of business and government are staffed by men and women who spend killing hours looking after the economic welfare and national security of the rest of us.

The rewards of this life are said to be very few: the love of money would be demeaning and, anyway, taxes are said to take most of it; its sacrifices are many, for failure brings economic depression to the masses or gains for communism as well as disgrace to the erring managers. Even the essential high-mindedness or altruism of our managers earns no reward, for the public is distracted, fickle, and, on occasion, vengeful. (The extensive literature on the "ordeal" of Lyndon Johnson is a case in point.) Hence for these "real revolutionaries of our time," as Walt Rostow has called them, self-fulfillment through work and discipline the only reward. The managerial process is seen as an expression of the vital personalities of our leaders and the right to it an inalienable right of the national elite. In addition to all of this, their lonely and unrewarding eminence in the face of crushing responsibility, etc., tends to create an air of mystification around technology's managers.

When the august mystery of science and the perquisites of high office are added to their halos, they glow very blindingly indeed. Thus, in ideology as well as in reality and appearance, the experiences of the higher managers tend to separate and isolate themselves from those of the managed. Again the situation within the US is not so severe nor so stark as in the Army in Vietnam but the effect on those who are excluded from self-management is very similar. Soldiers in Vietnam are not alone in believing huge, secret guns threaten them from various points; that same feeling is a national malady in the US. It seems fundamental to the social organization of modern technology that the quality of the social experience of the lower orders of society declines as the level of technology grows no less than does their literacy.

And, of course, this process feeds on itself, for with the consequent decline in the real effectiveness and usefulness of local and other forms of organization open to easy and direct popular influence their vitality declines still further, and the cycle is repeated. The normal life of men and women in the lower and, middle levels of American society now seems cut off from those experiences in which near social means and distant social ends are balanced and rebalanced, adjusted and readjusted. But it is from such widespread experience with effective balancing and adjusting that social rationality derives. To the degree that it is lacking, social irrationality becomes the norm, and social paranoia a recurring phenomenon. Those who seek an explanation for the infatuation of local government with anti-fluoridation campaigns several years ago need look no further. A similar irrationality is now being exhibited toward the war in Vietnam and the anti-war Movement.

With no great effort and using no great skill, Presidents Johnson and Nixon have managed to direct disorganized popular frustration over the continuation of the war and popular abhorrence over its unremitting violence on to precisely that element in the population most actively and effectively opposed to the war and its violence. As for paranoia, consider the widespread reaction of whites to the murder of Dr. King. Their demand for force and more force to be used against the Black population was consistent only with the hypothesis that Dr. King murdered James Earl Ray, just as SNCC members had lynched Klansmen only a few years before. People often say that America is a sick society when what they really mean is that it has lots of sick individuals. But they were right the first time: the society is so sick that individual efforts to right it and individual rationality come to be expressed in fundamentally sick ways.

Like the soldiers in Vietnam, we try to avoid atrophy of our social intelligence only to be led into fantasy and, often, violence. It is a good thing to want the war in Vietnam over for, as everyone now recognizes, it hurts us almost as much as the Vietnamese who are its intended victims. But for many segments of our population, especially those cut off from political expression because of their own social disorganization, the rationality of various alternatives for ending the war is fundamentally obscure. Thus their commendable desire to end the war is expressed in what they believe is the clearest and most certain alternative: use the bomb! Mesthene himself recognizes that such "negative externalities" are on the increase. His list includes "...pollution of the environment, potential damage to the ecology of the planet, occupational and social dislocations, threats to the privacy and political significance of the individual, social and psychological malaise...."

Minor matters all, however, when compared to the marvelous opportunities *laissez innover* holds out to us: more GNP, continued free world leadership, supersonic transports, urban renewal on a regional basis, institutional innovation, and the millennial promises of his school. This brings us finally to the ideologies and doctrines of technology and their relation to what I have argued is a growing gap in political culture between the lower and upper classes in American society. Even more fundamentally than the principles of *laissez innover* and the altruistic bureaucrat, technology in its very definition as the organization of knowledge for practical purposes assumes that the primary and really creative role in the social processes consequent on technological change is reserved for scientific and technical elite, the elite which presumably discovers and organizes that knowledge.

But if the scientific and technical elite and their indispensable managerial cronies are the really creative (and hardworking and altruistic) element in American society, what is this but to say that the common mass of men are essentially drags on the social weal? This is precisely the implication which is drawn by the *laissez Innover School*. Consider the following quotations from an article which appeared in *The New Republic* in December, 1967, written by Zbigniew Brzezinski, one of the intellectual leaders of the school. Brzezinski is describing a nightmare which he calls the "technetronic society" (the word like the concept is a pastiche of technology and electronics). This society will be characterized, he argues, by the application of "...the principle of equal opportunity for all but...special opportunity for the singularly talented few."

It will thus combine "...continued *respect* for the popular will with an increasing *role* in the key decision-making institutions of individuals with special intellectual and scientific attainments." (italics added.) Naturally, "The educational and social systems will make it increasingly attractive and easy for those meritocratic few to develop to the fullest of their special potential." However, while it will be "...necessary to require everyone at a sufficiently responsible post to take, say, two years of scientific and technical retraining every ten years...," the rest of us can develop a new "...interest in the cultural and humanistic aspects of life, *in addition to purely hedonistic preoccupations.*" (Italics added.) The latter, he is careful to point out, "would serve as a social valve, reducing tensions and political frustration."

Is it not fair to ask how much *respect* we carefree pleasure lovers and culture consumers will get from the hard-working bureaucrats, going to night school two years in every ten, while working like beavers in the "key decision-making institutions"? The altruism of our bureaucrats has a heavy load to bear. Stripped of their euphemisms these are simply arguments which enhance the social legitimacy of the interests of new technical and scientific elites and detract from the interests of the rest of us; that is to say, if we can even formulate those interests, blinded as we will be by the mad pursuit of pleasures (and innovation??!) heaped up for us by advanced technology. Mesthene and his schoolfellows try to argue around their own derogation of the democratic ethos by frequent references, as we have seen, to their own fealty to it. But it is instructive in this regard to note that they tend, with Brzezinski, to find the real substance of the democratic ethos in the principle of the equality of opportunity. Before we applaud, however, we ought to examine the role which that principle plays within the framework of the advanced technological society they propose.

As has already been made clear the *laissez Innover School* accepts as inevitable and desirable the centralizing tendencies of technology's social organization, and they accept as well the mystification which comes to surround the management process. Thus equality of opportunity, as they understand it, has precious little to do with creating a more egalitarian society. On the contrary, it functions as an indispensable feature of the highly stratified society they envision for the future. For in their society of meritocratic hierarchy, equality of opportunity assures that talented young meritocrats (the word is no uglier than the social system it refers to) will be able to climb into the "key decision-making" slots reserved for trained talent, and thus generate the success of the new society, and its cohesion against popular "tensions and political frustration."

The structures which formerly guaranteed the rule of wealth, age, and family will not be destroyed (or at least not totally so). They will be firmed up and rationalized by the perpetual addition of trained (and, of course, acculturated) talent. In technologically advanced societies, equality of opportunity functions as a hierarchical principle, in opposition to the egalitarian social goals it pretends to serve. To the extent that it has already become the kind of "equality" we seek to institute in our society, it is one of the main factors contributing to the widening gap between the cultures of upper and lower class America.

Approximately a century ago the philosophy of *laissez faire* began its period of hegemony in American life. Its success in achieving that hegemony clearly had less to do with its merits as a summary statement of economic truth than with its role in the social struggle of the time. It helped to identify the interests of the institutions of entrepreneurial capitalism for the social classes which dominated them and profited from them. Equally, it sketched in bold strokes the outlines of a society within which the legitimate interests of all could supposedly be served only by systematic deference to the interests of entrepreneurial capitalists, their institutions, and their social allies. In short, the primary significance of *laissez faire* lay in its role as ideology, as the cultural or intellectual expression of the interests of a class.

Something like the same thing must be said of *laissez innover*. As a summary statement of the relationship between social and technological change it obscures far more than it clarifies, but that is often the function and genius of ideologues. *Laissez innover* is now the premier ideology of the technological impulse in American society, which is to say, of the institutions which monopolize and profit from advanced technology and of the social classes which find in the free exploitation of *their* technology the most likely guarantee of their power, status, and wealth.

This said, it is important to stress both the significance and limitations of what has in fact been said. Here Mesthene's distinction between the positive opportunities and negative "externalities" inherent in technological change is pivotal; for everything else which follows inferentially from the actual social meaning of that distinction. As my analysis of the Vietnam bombing program suggested, those technological effects which are sought after as positive opportunities and those which are dismissed as negative externalities are decisively influenced by the fact that this distinction between positive and negative within advanced technological organizations tends to be made among the planners and managers themselves.

Within these groups there are, as was pointed out, extremely powerful organizational, hierarchical, doctrinal, and other "*technical*" factors, which tend by design to filter out "irrational" demands from below, substituting for them the "rational" demands of technology itself. As a result, technological rationality is as socially neutral today as market rationality was a century ago. Turning from the inner social logic of advanced technological organizations and systems to their larger social effect, we can observe a significant convergence.

For both the social tendency of technology and the ideology (or rhetoric) of the *laissez innover* school converge to encourage a political and cultural gap between the upper and lower ends of American society. As I have pointed out, these can now be characterized as those who manage and those who are managed by advanced technological systems. This analysis lends some weight (though perhaps no more than that) to a number of wide-ranging and unorthodox conclusions about American society today and the directions in which it is tending. It may be useful to sketch out the most important of those conclusions in the form of a set of linked hypotheses, not only to clarify what appear to be the latent tendencies of America's advanced technological society but also to provide more useful guides to the investigation of the technological impulse than those offered by the obscurantism and abstractions of the school of *laissez innover*.

First, and most important, technology should be considered as an institutional system, not more and certainly not less. Mesthene's definition of the subject is inadequate, for it obscures the systematic and decisive social changes, especially their political and cultural tendencies, which follow the widespread application of advanced technological systems. At the same time, technology is less than a social system *per se*, though it has many elements of a social system, viz., an elite, a group of linked institutions, an ethos, and so forth. Perhaps the best summary statement of the case resides in an analogy—with all the vagueness and precision attendant on such things: today's technology stands in relation to today's capitalism as, a century ago, the latter stood to the free market capitalism of the time.

The analogy suggests, accurately enough I believe, the likelihood that the institutional links and shared interests among the larger corporations, the federal government, especially its military sector, the multiversity and the foundations, will grow rather than decline. It suggests further a growing entanglement of their elites, probably in the neo-corporations of technology, such as urban development corporations and institutes for defense analysis, whose importance seems likely to increase markedly in the future. Finally, it suggests a growing convergence in the ethos and ideology of technology's leading classes along lines which would diminish slightly the relative importance of rhetoric about "property" and even about "national security," while enhancing the rhetoric of *laissez innover*. This does not necessarily imply and sacrifice in the prerogatives of either the private sector or of the crisis managers and the military, for one can readily understand how the elite strictures of *laissez innover* may be applied to strengthen the position of the corporate and military establishments.

A word about the elites: a number of writers of the *laissez innover* school, for example, J.K. Galbraith in his *The New Industrial State*, have argued that the enhanced importance of scientific and technical knowledge within advanced technological systems implies an enhanced socio-political power for the people who have such knowledge. Galbraith in particular has argued that these people, whom he calls "the educational and scientific estate," now constitute an elite class whose interests diverge quite sharply from those of other elites. Since I have argued at length elsewhere against this view, let me limit myself here to only a few critical observances. Galbraith's concept of an "educational and scientific" elite class overlooks the peculiar relationship which the members of that supposed class have to advanced technology. Specifically, it overlooks the fact that most technical, scientific, and educational people are employed at relatively specialized tasks within very large organizations whose managing and planning levels are hardly less insulated from their influence than from the influence of the technically unskilled.

The obvious growth in status and, I think, power of such men as Ithiel de Sola Pool, Herman Kahn, Samuel Huntington, Daniel Patrick Moynihan, Henry Kissinger, Charles Hitch, and Paul Samuelson hardly represents the triumph of wisdom over power—an implication not absent from Galbraith's analysis. An examination of the role which these men now play in our national life should emphasize that they are scientific and technical entrepreneurs whose power is largely based on their ability to mobilize *organized* intellectual, scientific, and technical manpower and other resources, including foundation grants and university sponsorship, in behalf of the objectives of going institutions. They are much more like managers than like intellectuals, much more like brokers than like analysts.

The same hardly applies to the thousands of E.E.'s, Ph.D.'s, M.S.'s, and so forth who make up the resources over which these mandarins preside and whose skills are so much at their disposal. But in recognizing the managerial and brokerage functions which predominate in the mandarin role we must also recognize that mandarin interests are more likely than not to converge toward the interests of other segments of the national elite. This conclusion is diametrically opposed to the one argued by Galbraith. A second major hypothesis would argue that the most important dimension of advanced technological institutions is the social one, that is, the institutions are agencies of highly centralized and intensive social control. Technology conquers nature, as the saying goes. But to do so it must first conquer man.

More precisely, it demands a very high degree of control over the training, mobility, and skills of the work force. The absence (or decline) of direct controls or of coercion should not serve to obscure from our view the reality and intensity of the social controls which are employed (such as the internalized belief in equality of opportunity, indebtedness through credit, advertising, selective service channeling, and so on). Advanced technology has created a vast increase in occupational specialties, many of them requiring many, many years of highly specialized training. It must motivate this training. It has made ever more complex and "rational" the ways in which these occupational specialties are combined in our economic and social life. It must win passivity and obedience to this complex activity.

Formerly, technical rationality had been employed only to organize the production of rather simple physical objects, for example, aerial bombs. Now technical rationality is increasingly employed to organize all of the processes necessary to the utilization of physical objects, such as bombing systems, maintenance, intelligence and supply systems. For this reason it seems a mistake to argue that we are in a "post-industrial" age, a concept favored by the *laissez Innover School*. On the contrary, the rapid spread of technical rationality into organizational and economic life and, hence, into social life is more aptly described as a second and much more intensive phase of the industrial revolution. One might reasonably suspect that it will create analogous social problems.

Accordingly, a third major hypothesis would argue that there are very profound social antagonisms or contradictions not less sharp or fundamental than those ascribed by Marx to the development of nineteenth-century industrial society. The general form of the contradictions might be described as follows: a society characterized by the employment of advanced technology requires an ever more socially disciplined population, yet retains an ever declining capacity to enforce the required discipline. One may readily describe four specific forms of the same general contradiction. Occupationally, the work force must be over-trained and under-utilized. Here, again, an analogy to classical industrial practice serves to shorten and simplify the explanation.

As a device in the organization of the work process the assembly line is valuable mainly in that it gives management a high degree of control over the pace of the work and, more to the point in the present case, it divides the work process into units so simple that the quality of the work performed is readily predictable. That is, since each operation uses only a small fraction of a worker's skill, there is a very great likelihood that the operation will be performed in a minimally acceptable way. Alternately, if each operation taxed the worker's skill there would be frequent errors in the operation, frequent disturbance of the work flow, and a thoroughly unpredictable quality to the end product. The assembly line also introduces standardization in work skills and thus makes for a high degree of interchangeability among the work force.

For analogous reasons the work force in advanced technological systems must be relatively over-trained or, what is the same thing, its skills relatively under-used. My impression is that this is no less true now of sociologists than of welders, of engineers than of assemblers. The contradiction emerges when we recognize that technological progress requires a continuous increase in the skill levels of its work force, skill levels which frequently embody a fairly rich scientific and technical training, while at the same time the advance of technical rationality in work organization means that those skills will be less and less fully used.

Economically, there is a parallel process at work. It is commonly observed that the work force within technologically advanced organizations is asked to work not less hard but more so. This is particularly true for those with advanced training and skills. Brzezinski's conjecture that technical specialists undergo continuous retraining is off the mark only in that it assumes such retraining only for a managing elite. To get people to work harder requires growing incentives. Yet the prosperity which is assumed in a technologically advanced society erodes the value of economic incentives (while of course, the values of craftsmanship are "irrational"). Salary and wage increases and the goods they purchase lose their over-riding importance once necessities, creature comforts, and an ample supply of luxuries are assured. As if in confirmation of this point, *Fortune* has pointed out (January, 1969) that among young people one can already observe a radical weakening in the power of such incentives as money, status, and authority.

Politically, the advance of technology tends to concentrate authority within its managing groups in the ways I have described. But at the same time the increasing skill and educational levels of the population create latent capacities for self-management in the work place and in society. This aspect of the contradictions inherent in technology seems especially noteworthy in much of the current dissent within the armed forces. Of course, there has always been griping in the Army, but the fact that the griping now attaches it to political problems—the war, the rights of servicemen, and so on—clearly speaks to the fact that GI educational levels have increased very radically. A similar explanation casts light on the campus revolt. As Lionel Trilling has pointed out, the cultural and intellectual level of today's university students is far higher than that of their predecessors, if only because of television and the fact that a good part of traditional college work is now completed in the better high schools.

The claim to greater self-management by university students is the natural outcome of this change. At the same time, however, the university has been developing the power and status of its own elites. These include the research elites whose power and status are based on consultant ships, on their ability to win research grants and contracts and thus prestige for their universities, as well as the more general category of professional elite, i.e., those whose power within the university is buttressed by their prestige in the national professional associations and expressed in their control of the major academic departments. In spite of the apparent decentralization within the university organization, one must recognize that there has been a very tangible increase in the influence of these elites. Within the university (as elsewhere) power has become more centralized, that is, it has gravitated toward an identifiable collection of research and professional mandarins. It is the power of the latter over curriculum, admissions, research and consulting policy, hiring and advancement that is being challenged by student dissidence.

It should be added here that even the "irrationality" of the student revolt is clarified by the general lines of this explanation, i.e., its common failure to formulate coherent social programs and its tendency to enter battle under extremely vague symbolic banners. As we have seen social irrationality can be explained as a normal effect of the social and organizational patterns of advanced technological systems and, if anything, is increased by the personal intelligence of the people trapped in those systems. Finally, there is a profound social contradiction between the highly stratified society implicit in, say, Brzezinski's meritocracy and the spread of educational opportunity. Yet each appears equally required by advanced technology. These are brief and, I believe, barely adequate reviews of extremely complex hypotheses. But, in outline, each of these contradictions appears to bear on roughly the same group of the American population, a technological underclass. If we assume this to be the case, a fourth hypothesis would follow, namely that technology is creating the basis for new and sharp class conflict in our society.

That is, technology is creating its own working and managing classes just as earlier industrialization created its working and owning classes. Perhaps this suggests a return to the kind of class-based politics which characterized the US in the last quarter of the nineteenth century, rather than the somewhat more ambiguous politics which was a feature of the second quarter of this century. This leads to a final hypothesis, namely that *laissez innover* should be frankly recognized as a conservative or right-wing ideology This is an extremely complex subject for the hypothesis must confront the very difficult fact that the intellectual genesis of *laissez innover* is traceable much more to leftist and socialist theorizing on the wonders of technical rationality and social planning than it is to the blood politics of a De Maistre or the traditionalism of a Burke.

So be it. Much more important is the fact that *laissez innover* is now the most powerful and influential statement of the demands and program of the technological impulse in our society, an impulse rooted in its most powerful institutions. More than any other statement, it succeeds in identifying and rationalizing the interests of the most authoritarian elites within this country, and the expansionism of their policies overseas. Truly it is no accident that the leading figures of *laissez innover*, the Rostows, Kahn, Huntington, Brzezinski, to name but a few, are among the most unreconstructed cold warriors in American intellectual life. The point of this final hypothesis is not primarily to re-impress the language of European politics on the American scene. Rather it is to summarize the fact that many of the forces in American life hostile to the democratic ethos have enrolled under the banner of *laissez innover*.

Merely to grasp this is already to take the first step toward a politics of radical reconstruction and against the malaise, irrationality, powerlessness, and official violence that characterize American life today. The most important thing to notice about the articles "The Role of Technology in Society" and "Technology: The Opiate of the Intellectuals" is that they provide an *explicit debate*. This debate revolves around differing conceptions of the place of technology in society. When examining a debate, it is important to notice where the debaters agree and where they disagree. In the introduction, Teich obscures the areas of agreement by informing the reader that the authors have a fundamentally different perspectives based on the vantage point of the powerful and the powerless. That's true but it may not be the best way to categorize the differences and it doesn't take into account some very interesting similarities.

Points of Agreement:

1. Mesthene and McDermott both agree that technological positivism and pessimism really miss the complex and subtle interactions between technological development and society. Both writers want to explore those "subtle relationships".
2. Mesthene and McDermott both agree that technological advance has *negative* consequences that relate directly to technological change. Mesthene is not a simple or simplistic cheerleader for new technologies.
3. Mesthene and McDermott both criticize the presumed *mystical autonomy* "presumed to lie in technology" and demand that technological change and social values be better aligned.
4. Both Mesthene and McDermott suggest that a major problem with the introduction of new technologies is social and individual *dislocation*. Mesthene, for example, points out that the development of new transportation systems widened the gap between the rich and poor and made the American inner cities places of poverty and unemployment.
5. Both Mesthene and McDermott agree that that one of the major problems of a technologically advancing capitalist nation the irresponsibility of private individuals and large corporations who put profits before the interests of people.
6. Mesthene and McDermott would also agree that one of the most serious negative effects of technology is the erosion of traditional freedoms and democratic processes that Americans fought for over the course of two centuries.

7. Both Mesthene and McDermott conclude that the scope and complexity of "public decision making" in a technological society results in *systems of social organization* that are centralized, controlling and fundamentally undemocratic.
8. Both Mesthene and McDermott believe that, while the overall level of social control is increasing dramatically, it has become much more subtle, indirect and manipulative.
9. Both Mesthene and McDermott would probably agree that we need "new institutional farms and new social mechanisms" or the "political conflict" of the 1960s will increase.

Laissez Innovers

There appear to be so many points of agreement between Mesthene and McDermott that you might wonder why McDermott is so aggressive in his attack on what he calls *laissez innovers*. Let's define the term to see what McDermott is getting at. *Laissez-faire* means 'let it be'; the term was coined by eighteenth-century French economists do define a free market in grain. Gradually, it came to be equated with the creation of a free market (i.e. capitalist economy). McDermott has serious reservations about a free market because he believes that capitalist economics means the exploitation of the powerless by the powerful for their own ends. In other words, the term *laissez-faire* was not neutral. It involved elements of control and oppression that people had to fight against in order to create a more humane and democratic society.

To a certain extent, McDermott believes that the American people did win significant rights in the twentieth century through their greater involvement in the political process. McDermott believes that the people are losing their ability to influence the political process and to defend their interests in a developing technological society. *Laissez innover* means freedom for technological innovation. The lack of significant social controls over technology means those who control the direction of technology have much more power over people's lives than ever before. In order for *laissez innover* to exist, social relations and social values have to be *rationalized* in ways that allow the controllers of technology to take advantage of the opportunities that technology provides. The primary emphasis in a society based on *laissez innover*, therefore, is constant technological innovation that is mirrored and reinforced by institutional innovation.

In such a technology-directed environment, McDermott suggests, human values must be directed by technological development. But technological development is not simply an *abstraction*. Technology is a series of innovations, and also systematic processes, that are directed by a small group of people. These elites push technology in directions that are favourable to the self-interest of their group. In a technological society, they are able to do this very effectively because the vast majority of people haven't got the specialized expertise or the understanding of highly complex systems that would allow them to participate in decision-making.

The Technocratic Perspective

Mesthene clearly understands the problems and disadvantages associated with the introduction of new technologies. He is also willing to criticize capitalists and corporations who don't seem to understand the damage and social dislocation that new technologies can bring in their wake. None of this, however, stops him from being a *technocrat*. A technocrat is someone who advocates a government or social system that is controlled by scientific technicians. At no point in his article does Mesthene ever suggest that technology should be controlled by a democratic society. Quite the reverse; he believes that "decision-making structures" need to be *adjusted* "to the realities of technology so as to take maximum advantage of the opportunities it offers and so that we can act to contain its potential ill effects before they become so pervasive and urgent as to seem uncontrollable."

Who is going to achieve this balancing act? Mesthene doesn't say explicitly, but he does give us lots of information that allows us to draw the appropriate conclusion. It has to be achieved from *above,* since most of the people don't understand the complexities of a modern technological society. It can't be achieved, or if so only *formally,* by represented officials, because most of them don't have the specialized skills in data collection and systems analysis that are required for the task. The groups who will marry the needs of a complex society are trained *experts* who can commit the enormous amount of time and develop the skills necessary to "foster technology on a large scale."

Mesthene suggests an enormously important role for two groups in society. The first are the technological innovators - scientists and technicians - who generate *opportunities.* The second are the *social engineers* or those whose job it is to make sure that the political structures and values of society conform to the reality of a modern technological society. The net result is a society controlled by experts with a technological agenda. The average person needs to have their values changed to come into line with technological reality. Popular democracy is a thing of the past; society now must be rationally managed.

Is this assessment of Mesthene too harsh? Look at what he says. He says that the entire *logic* of modern decision making "requires greater and greater dependence on the collection and analysis of data and on the use of technological devices and scientific techniques." In other words, a technological society will be created by fundamentally technological means. Second, he says that public policy can no longer be based on anything other than *reliable knowledge* which he describes as "a strong capability to collect, aggregate, and analyze detailed data about economic activities, social patterns, popular attitudes, and political trends." Third, he suggests that we as individuals can no longer play a significant role in decision making because modern societies are *complex systems* rather than collections of free and independent individuals. In a complex system, everyone is dependent on everyone else and individuals need to be controlled to ensure the greatest health of the social organism.

Fourth, he dismisses any defense of the *dignity of the individual* or objection to our "increasing dependence of decision making on scientific and technological devices and techniques" with the claim that a small group of knowledgeable experts needs to "save the society before it explores under plan less and inadequate administration. In a technocratic world, "social and psychological displacements" are not signs that the system itself may be haywire. Rather, they are problems or "paradoxes" for technocratic experts to solve. Technology is the *motor* that continually runs ahead of man's ability to deal with it.

Technology as Opiate

In "Technology: The Opiate of the Intellectuals", John McDermott claims that this technocratic position gives technology way to much power over people. It transforms technology into an abstraction. That abstraction, despite the acceptance of *negative* problems associated with technology, is inevitable. It is also *scientifically pure,* just as long as it is controlled by experts - intellectuals like Mesthene. With the intellectuals at the help, technology can be a *self-correcting system.*

McDermott thinks that this viewpoint is extremely naïve. Take for example, Mesthene's *anti capitalist* statements. McDermott suggests that Mesthene's heart might be in the right place; Mesthene might think that he and his friends will shape technology for the benefit of mankind. But what guarantee is there that the new technocratic elite will not simply perpetuate themselves and increase their power in society? Even worse, what it these technocratic elite merely increase the power of those who really control the production, use and distribution of new technologies - the people that McDermott stereotypes as profit driven corporations and their "managerial cronies."

For McDermott, a technocratic society is not an abstraction but a specific set of social relationships in which the power to make decisions in concentrated in the hands of scientific-technical bureaucracies. Technocracy is not so much about finding ways to bring social institutions into line with technological realities. It is an institutional movement and culture in its own right that "improves and rationalizes the performance of its machines and men" within quite specific and tightly "integrated systems." These systems place such an emphasis on the control of any "negative externalities" that they are remarkably resistant to "intervention by persons or problems operating outside or below their managing groups."

In a technocratic scenario, politics becomes irrelevant. There is no need (a la Mesthene) to bring politics into line with technology because "technology creates its own politics." Not only is the natural impulse of technology to create highly efficient systems that limit the potential for individual disruption, but also they are so "capital intensive" that the power of the individual is effectively inconsequential. Only those who have senior managerial power to control the flow of capital investment, or the technical knowledge to take advantage of opportunities, have any real power in a technocratic society.

For McDermott, the spread and mystification of technocratic values is a serious threat to democracy. The formal ingredients of democracy might persist for some time, but a technocratic society cuts the majority of people out of power sharing and limits their influence on the decisions made that profoundly affect their lives. A technological society that is controlled by scientific and technical elites takes knowledge out of the hands of the people and reserves it for a small group - the *meritocracy*.

The Rise of the Meritocracy

McDermott views technocracy as an attempt to centralize and control power in the hands of the few against the many. Someone like Mesthene could conceivably argue that, while some dislocation and perceived powerlessness is the inevitable result of technological change that does not mean that the fundamental requirement for a democratic society ceases to exist. He could, and probably would, argue that a technocratic society is democratic because it needs to identify and reward the smartest and most hard working people in society - the group that McDermott dismisses as "the intellectuals." To be sure, not everyone can be a member of the elite. But assuming that Mesthene is right and we can create an elite cadre of scientists and technicians from our university and college system, would that be such a bad thing? Just as long as everyone has "equality of opportunity", the best people would rise to the top.

Isn't that what everyone wants a democratic society to do? Isn't one of the problems with more primitive capitalism precisely that it distributes social rewards unevenly and sometimes unfairly. Many technocrats like Mesthene believe in the power of higher education to ensure that people with *merit* rise to the top. Mesthene himself is an example of a person who has *achieved* distinction at one of America's top universities - Harvard. On the other hand, McDermott appears to have had a much more uneven experience in the Department of Labour Studies at the State University of New York. Perhaps, McDermott's criticism of Mesthene is partly the result of sour grapes and the "scientific and technical elite and their indispensable managerial cronies are the really creative (and hardworking and altruistic elements in American society."

McDermott doesn't really address the issue of *merit* head on. But we can illuminate the debate somewhat by doing it for him. He could suggest, for example, that a technocratic society favours those who come from families with the economic means to pursue higher education. He might want to argue that a *meritocracy* cannot exist without a massive investment in education generally, in order to make sure that those with ability have the opportunity to achieve. He could point out that many so-called *meritocracies* do not really promote people with the best ability; in fact, they tend to be self-perpetuating and to provide a few people with quite exceptional opportunities to achieve. These are questions that go to the very heart of our present system of higher education and give rise to intense differences of opinion.

While McDermott hints at some of these issues, however, they are not his primary concern. His real issue is that a technocratic society, a meritocracy if you will, relegates the majority of the people to minor roles in society. McDermott puts this rather strongly when he suggests, "The common mass of men is essentially drags on the social weal." Thus, technocracy breeds an elitism that strikes at the heart of democracy. That kind of elitism is only implicit in Mesthene's discussion of the social displacement that comes with technological development. But in other writers, it can be quite explicit. Zbigniew Brzezinski, writing ironically for the left leaning *The New Republic*, suggested that the *technetronic society* of the future would be run by a "singularly talented few". The rest of the population would have to content themselves with being passive players in society and pursuing their own individual interests.

The Imperative of Control

Equality of opportunity in a technocratic society, paradoxically, creates a very inegalitarian, hierarchical, and role stratified society. In the past, such societies have been inherently unstable and characterized by class or other kinds of conflict. But conflict is the antithesis of a technocratic society, which requires a high degree of social control in order to ensure the integrity of systems. It is the implicit element of social control that most worries McDermott. Both McDermott and Mesthene agree that a technological society means centralization and the control of power in the hands of a few. How will the many be controlled so as not to threaten the few? Mesthene provides some suggestions when he claims that the increased knowledge and subtle techniques of the key decision makers will allow them to develop new institutional forms and new mechanisms for better *integrating* individuals into the system.

Even if these techniques were not explicitly coercive, it is difficult to see how they could avoid being manipulative. After all, the system cannot function unless discordant negative externalities are suppressed. McDermott mentions other forms of social control, the chief being the ideology of technology itself. Technology may have conquered nature, but it can only conquer man if people internalize it. That is why the advocates of technology constantly push not only the rationality of technology, but also its millenarian promise.

A technological society cannot rely on communicating a future utopia in the face of problems and dislocations, so it also promotes a *hedonistic lifestyle* that distracts people from their powerlessness. Thus Brezinski, for example, is quite open in suggesting that *hedonistic preoccupations* will "serve as a social valve, reducing tensions and political frustration." Political men and women are transformed into a consumer community, aided by advertising, the promotion of credit cards, and selective service channeling (i.e. cable television). Writing in the 1960s, McDermott believed that the technological system was far too complex, and its controls too indirect, to be stable. He believed that "advancing technology" had "an ever declining capacity to enforce the required discipline..." Of course, he was writing during the hippy era, when the civil rights movement and anti-war protest were at their height. During such a period, it was possible to envision alternatives to the status quo and even to posit social revolution. The technological society has advanced since then.

The Altruistic Bureaucrat

Clearly, despite their areas of agreement, McDermott and Mesthene have distinctly different views of technology. For Mesthene, a technological society is inevitable and provides humanity with opportunities. The trick is to deal with the problems; something that can only be achieved if power is given to a small group of people to ensure that scientific and social engineering are in harmony. These people would by definition be modern society's *problem solvers* and entirely neutral because they are rewarded for their disinterested ability to bring their scientific and technical knowledge to bear on social problems. For McDermott, this kind of thinking borders on technocracy. A technocratic society is not desirable because it would mean the end of democracy.

Moreover, it would be inherently unstable and conflict ridden because it deprives the majority of any effective power. Where McDermott and Mesthene disagree most violently is in their assessment of the new scientifically trained elite that would have hegemony in a technological society. For Mesthene, these individuals are the *saviours* who would "save the society before it explodes under planlessness and inadequate administration. For Mesthene, this would simply represent new self-perpetuating social elite. By presenting himself and his colleagues as anything more than right wing defenders of the status quo, Mesthene is either being totally naïve or engaging in deception for reasons of self-interest.

One aspect of this debate is worth considering in our present context. Mesthene's argument is mildly anticapitalist. The elite that Mesthene describes could come from the ranks of business, but his heroes more closely resemble scientists, technicians, and civil servants (the *social engineers*). When Mesthene wrote his article, university scientists were much more independent than they are today, when universities and governments are looking at the bottom line and increasingly seeking partnerships with the corporate community. Moreover, Mesthene wrote his article at a time when the public sector in North America was still at the height of its power, prior to the advent of privatization and the exodus of some of the public sector's finest minds.

Do the changes of the past 4 decades strengthen or weaken Mesthene's or McDermott's arguments? Consider that the corporate community has much greater power over government and the direction of technology than it did in the 60s. Consider also that the social conflicts of McDermott's time have dissipated. Does putting their arguments in historical context make these arguments as relevant today as they were when they were written?

5

Evaluate Technology

While the objective of technology transfer is to transmit advantages to its acquirer or licensee, transferring technology can involve serious risks. The terms "transport" and "transplantation" of technology appear often in this module to underscore the feasibility of successfully transferring the technical attributes of a technology which make it excellent and competitive – especially to developing countries. For example, it may or may not be feasible to transport automation, which may be viewed as a technical attribute, to a country where industry is characterized by a relatively low level of technological complexity.

The term "technology" has many connotations. In this module it connotes industrial technologies, i.e. the technical means employed for producing established commercial goods. Technology can be a package of various kinds of information (general, specialized or proprietary) coupled with the technical, managerial and craft skills embodied in an individual team. Depending on the intended product's nature and capacity, the package can be simple or complex. Viewed differently, technology is a combination of invention, scientific discovery and principles, data accumulation and experience in constructing, using and servicing equipment and machinery.

Good technology is configured to obtain a desired commercial result economically, conveniently and safely and to ensure its maintainability. It may be supported by patents, trade marks, copyrights or proprietary data and information (intellectual property rights), which enhance its commercial value. Acquiring technology bypasses the rigorous of the trial-and-error process and yields immediate, significant rewards for the buyer.

By and large, industrial technologies belong to one, or a combination, of the following five categories:

- Technologies for the production of goods, measured, or classes of goods, of near identical design or properties, whose output is in volumetric units. Technologies oriented to the manufacture/production of ores, implements and tools, glass sheets, cement, caustic soda, vegetable oils, pharmaceuticals, machinery (e.g. lathes), or components of machines and appliances typify this category. The production operation itself may involve mining, extraction, fabrication, assembly formulation, physical transformation or chemical conversion (process).
- Technologies that enhance the properties, features or qualities of a product to create a commercially relevant advantage such as cost, convenience, performance or safety. Examples of product improvements include "free-flowing" salt, the twin blade shaving razor, cocoa butter substitutes for making chocolates and encapsulation of pharmaceuticals.
- Technologies that produce one or more special effects, e.g. Hot and cold rolling steel; anodizing, galvanizing or electroplating metals; texturizing yarn; waterproofing, fireproofing and dyeing fabrics; fumigating and detoxifying grain, tubers or feeds; sterilizing and pasteurizing dairy or other processed food products. The product may be produced in-house, be a bought item, or it may be contracted out to a third party with expertise in the specific conditioning process.
- Technologies that modify a production process or manufacturing system to bring about some advantage

or leverage. Examples include: improving operating or public safety by eliminating pollutants in products and processes, removing phosphates in detergents and cleaning up automobile exhaust gases. Such technologies can also comprise the automation, computerization or robotization of processes and mechanical sequences to improve a particular feature. (These technologies do not always have a microeconomic significance but can enhance the image of the technology).

- Technologies for the "production" of technical services, e.g. Product/process design and engineering, computer software development and modeling and productivity enhancement techniques.

Since developing countries tend to be interested in technologies in the first three categories, which can stimulate entrepreneurship or economic development, this module focuses on the analysis and evaluation of those types. For the following analysis, it would be helpful to consider the implications of product and technology life cycles for the exercise of technology evaluation. The curve of product life cycle is S-shaped and shows three phases: growth, ascent and maturity phases. By then, however, they will reflect the growth of the industry rather than of the product.

Overtime, some earlier forms of a product phase out (fade into oblivion) or shift from one industry to another. For significant innovations like the automobile, television and computers, for which the overall product life cycle is comparatively long, the sub cycles constituting it may be of different lengths. The technology life cycle has four phases: latent development, ascent, maturity and decline. Very basic products have very long product lives, but particular technologies involved in making or servicing these products (typically) have shorter lives. The technology life cycle can be regarded as having a rising potency on the ascent phase; it reaches a peak at some point and declines until it ceases to have valid potency.

The Technology Transfer Framework

By and large, most technology originates in industrialized countries, although the newly industrializing countries are becoming important contributors. A technology package develops to meet an existing or forecast market need in its country of origin and is consistent with accessible resources. It can be said to be "appropriate" to that environment. Although such a technology may later be modified to use different raw materials and simpler levels of automation, it will still be recognizable as mature technology. A mature technology reflects the capabilities of mass production, the market preferences of an affluent population and the workability and efficiency of a well-developed industrial infrastructure. It also embodies a large number of major and minor improvements made over a long period of time. Furthermore, a mature technology reflects the strength imparted to it by the legal framework in which it is used.

A state-of-the-art technology, particularly one on the ascent portion of the technology life cycle, is unlikely to be available for licence, especially to a licensee in the relatively amorphous marketing and legal environments of a developing country. Even if it were available, it would most likely be inappropriate for a developing country, not only for the reason cited above but for others as well, such as a lack of technological complexity in the industrial structure of the host country. In industrialized countries technologies yield products or processes with specifications different from those of the innovative products. The differences result from the processes of competition and imitation and from the mechanisms of market segmentation, product positioning and niche markets, which both multiply technologies and broaden their range.

Companies often consider exporting technologies to subsidiaries, joint ventures, or third parties for a number of reasons: to avoid home-market competition, to enlarge international market share, to gain access to more conducive markets and to offset development costs. The technologies may be offered for licence directly by the owner of the technology or through engineering companies and licence brokers. For almost the same reasons, the technologies applied in products and processes in the mature or declining phase of the technology life cycle may become offered for licence and sale.

The Reconfiguration of Technology

It is generally difficult to transplant technology from one environment to another. In other words, products or processes developed in the context of one market environment are seldom wholly suitable in another. This sometimes holds true even for technologies transported between industrialized countries. For example, the composition and form of the leading detergent products are markedly different in the United States, Western Europe and Japan even though the principles under which detergents are formulated remain the same. This happens because of differences in washing traditions, the fibre mix in the laundry basket, demographic factors etc. The same phenomenon occurs in the motor car industry; even though the manufacturer and trade marks are the same in different countries, the car models differ significantly in such features as seating capacity, number of doors, thrust, acceleration, fuel consumption, suspension and comfort.

The transportability of technology to developing countries is affected by other factors as well, including small markets, raw materials constraints, scarcity of skills and underdeveloped infrastructure. Thus, except for the simplest transfers, technologies by and large will either need to be modified and made appropriate for the new environment or they will have to be accepted even though inappropriate. Transferring a particular technology to a developing country typically requires that it be modified in one or more of the following ways:

- Scaling down, so that it meets the requirements of the new marketplace, mainly reduced capacity and minimum penalties for lower levels of product quality and economic efficiency.
- Redesigning it to use scarce inputs in ratios that are economically rational in the new environment.
- Ensuring its maintainability and its ability to be absorbed at the skill levels available (or trainable in the new environment).

It may be necessary to restyle the products or to reengineer the production technology, or both. Modifications should be carried so as not to jeopardize the technology owner's intellectual property, trade marks, competitive standing or international image. However, a technology owner has no means of knowing, if say, a scaled-down version of it (or a version that, uses another raw material, or that has been simplified will work effectively or efficiently in the new environment except, perhaps, by attempting process simulation, "pilot-planting" or market-testing. Unless the costs of this testing can be passed on to the technology recipient, in whole or in part, the owner will have to absorb them. The transfer of technology to an environment different from that in which it was developed entails risk, so a methodology is needed to identify and appraise technology for its acceptability.

The Ideal Technology Selection Process

In an ideal approach to selecting technology appropriate to an environment. The entire acquisition and implementation process must be considered, not merely the financial and technical merits and risks. Technology must be viewed as a product exhibiting a degree of "stickiness" to its owner. That is, it reflects the perspective of the owner: his attitudes to economic, technical and other factors, and his desire to control the technology through legal means. When selecting a technology, one must not only evaluate its technical excellence but must also reckon with the firm that developed it: the quality of legal protection the firm has acquired its reputation for successfully operating technologies, and the extent of its international activities.

The tendency to turn to transnational corporations as sources of good technologies, or to corporations known for excellence in certain fields, e/g. fibres or audio equipment, demonstrates this stickiness factor. Consequently, the technologies that emerge may be equivalent in terms of product types and outputs, but very different in their use of raw materials, energy and other inputs, in their manufacturing and product specifications; and in their patents, trade marks and other proprietary rights. To a substantial extent, the stickiness factor indicates the degree of support one can expect from the developer in making a technology viable in its new home.

Although technology is one of the mot important factors of production, its value to a society cannot be characterized in the same manner as other factors of production, i.e., "interest rates" for capital, "rentals" for land, and "wage rates" for labour. Units of measure, such as the running royalty rate for licensed technology, do not necessarily allow comparability: just because one technology is offered at a higher royalty rate than another equivalent technology doe not mean there is an objective, qualitative difference that makes it superior.

Open and Closed Architecture Technologies

In this module, appropriateness and economic-technical risks are used as two key parameters for evaluating if a technology will be suitable in the host country environment. This is based in large part on responses obtained from the technology owner, a situation not unlike discussions between doctor and patient. On-site inspection of the working technology by its intended recipient, which would be available before making a decision, will generally not be possible until a degree of contractual certainty is created. Before proceeding to a discussion of evaluation tools, it may be useful to classify the technology differently than in the introduction (where five categories were given), that is into two broad categories: those with "open" and "closed" architectures. Doing so bears on the scope of analysis available in selecting technology.

Technologies that relate to assembling components to make a product such as a washing machine or lathe or that relate to making mature commodity products, such as cement, typify "open-architecture" technologies. In the case of an assembled product, a competent professional can actually disassemble the product to see how it has been put together. Such an examination permits determining which components are most critical to operating the appliance or machine and how effectively each performs relative to its counterparts in an equivalent appliance or machine. Likewise, a cement making process offered as an "engineering package", which would disclose its salient features, can be conceptually disassembled into its component elements. Using the wealth of information available in technical literature, the probable sequence of physical/chemical operations by which cement is manufactured in the engineering package can be visualized.

Technologies that have entered the public domain through the expiry of patents also belong to this category. Indeed, the first IBM personal computer was expressly, designed to have an open architecture so that industry would be able to manufacture peripherals (such as printers) and software, thereby expanding its usage. Such analyses can help a technology evaluator appreciate the excellence of the technologies offered. An evaluator can then develop inquiry procedures seeking clarifications and assurances from the technology owner in areas of importance, doubt and uncertainty, and on issues affecting the "relocating" that is, the transportability of a technology. In "closed-architecture" or "closed-system" technologies – such as those for manufacturing novel alloys, drugs, polymers, or integrated chips examining the end product provides little information about the raw materials used, the manufacturing process, the conditions during manufacture, the processing sequences involved etc.

The product or process cannot be conceptually disassembled, except in the vaguest terms. Practically all crucial aspects of the process must be is closed to the technology recipient for him to assess its appropriateness and risk. Thus, technologies with an open architecture are generally easier to assess because there is greater opportunity for examination prior to acquisition or licence than there is for closed-system technologies. Of course, many technologies are partially open architecture and partially closed-systems. Nevertheless, testing for appropriateness of technologies depends on obtaining some level of process disclosure from the technology owner.

The amount of material available for examination and the knowledge as well as the experience of the technology owner in applying the technology, reveal themselves only as the collaborative arrangement between the owner and potential technology-recipient gains strength. Even so, much of the technology's nature will remain unrevealed. It is not always possible for developing country entrepreneurs to go through the sequences to examine alternative technologies. In many cases, the choice is between accepting or not accepting a single offer of technology resulting. This can happen for several reasons including the following: (a) not knowing that other sources of technology exist, (b) lack of any other willing supplier of technology, or (c) the fact that the technology owner is assuring a market for the product.

Assessing the Appropriateness of a Technology

Assessing technological appropriateness involves assessing the technical and economic features of a technology package in the context of production in a given national environment. The assessment process requires some level of information disclosure from the technology owners, from obtaining responses to queries, to visiting plants of the licensor, to obtaining confidential disclosures (drawings, designs, specifications) and so forth.' By and large, it will be difficult for technology evaluators in most developing countries to obtain the needed information without providing the technology owner with some assurances. In some legal environments, prior disclosures and "look-see" arrangements may be obtained by paying front-end fees and the technology need not be selected. Typically, developing country Governments discourage such payments, although they are widely practiced in developed countries.

As a result, evaluation of appropriateness is carried out under less than ideal conditions. However, good homework by the potential acquirer of technology, striking a good relationship with the technology owner, demonstrating seriousness of purpose and sending strong signals that good technology will find a new and rewarding habitat; can stimulate responses useful enough to make good decisions. Where there is a choice of several technologies, analyzing appropriateness is much easier than analyzing feasibility for a single technology. A plurality of choices inherently shows that there are several accessible and practiced routes to achieve a given objective. It also shows that some technologies have facets that enable them to work in different habitats. Moreover, one route may have a configuration close to that required by the technology recipient.

However, as an initial exercise, it may help to analyse appropriateness by assuming only a single technology offer of a stand-alone technology not influenced by extraneous parameters such as financial credits, equity participation. In the following hypothetical cases, two of them with relatively open architectures and one of the closed system type, the first step is to develop checklists for evaluating the technologies.

Two Cases of Open-architecture Technology

In this case the product is one that can be easily disassembled by an engineering professional. It can be put together by obtaining from its manufacturer semi-knocked-down (SKD) or completely knocked-down (CKD) product kits. However, even though little "technology" is apparently needed to assemble the product from its parts, many things would not be known even to a professional. Several questions arise. How is the assembly best sequence? Which subassemblies are made first and which later? How fast can the assembly be done? Where are the hold ups? What kind of a floor layout is bet suited to assembly? What quality control measurements are made and what kinds of instruments are required? At what stages of product assembly are the sub-assemblies tested?

Would local technicians need to be trained? Thu, although we are dealing with what might be called "screwdriver technology", many things that should be obvious from the open architecture of the technology are, in fact, not. None the less, the questions raised above can be answered. They will form part of the "technology package" – the know how (or should it be called show how?) to be acquired from the proprietor of the technology.

In this case, a technology whose features are largely available in the public domain is tested for appropriateness. Its features are fairly well described in technical literature (including in expired patens) and can be explored through the use of consultants who have investigated or practiced similar technologies. It is assumed that there is an ongoing national market for the product that the entrepreneur can manage project finances and that he is capable of establishing the enterprise and organizing its operations. The technology at hand involves the manufacture of copper-based welding rods, used as a filler metal for joining ferrous and non-ferrous metals through braze welding with a gas torch.

A literature search and advice from consultants has disclosed that in a typical manufacturing scheme, virgin metals, eg. copper, zinc and tin, plus hardeners, if needed, such as phosphor-bronze, are melted, under flux cover, in graphite crucibles, and the molten metal is cast into rods on green sand moulds. The rods are then hot-rolled to reduce their diameter and then cold-rolled and annealed before being sent to wire-drawing machines from which the end product emerges after pickling. Further, annealing may be practiced for certain grades. The national market that meets supports American Welding Society (AWS) – American Society of welding rods.

The professional consultant is of the opinion that (a) most of the information on the manufacturing process is in the public domain (open architecture), (b) all of the manufacturing equipment involved can be sourced locally at competitive prices and (c) the local environment can accept a product of this complexity. These factors, by themselves, are insufficient to warrant successful entry into the market place. A helping hand is needed perhaps from a firm active in a similar market in another country and having a diversified range of products and a good product mix. In this case, show how is not as important as knowhow pertaining to the manufacture of a wide range of products.

A prospective buyer needs a basic knowledge of the operating process and a preliminary idea of what technical support will be necessary before he or the licensee can elicit enough information from a technology supplier to begin an evaluation of technology's appropriateness. A check-list has been developed of the kinds of questions a technology supplier might be expected to answer for a serious-minded client. The responses help the technology seeker to determine the basic features of the manufacturing process, assess the technical factors critical to commercial success, identify areas of technical risk and highlight matters that should be incorporated into the final transfer contract.

Many aspects of the technology and its appropriateness should become apparent from the responses to thee questions. "Look-see" arrangements may come next, possibly at a cost, since a visual check is often vital for technology selection. In neither case is there much need to probe risk elements, because no risk areas are evident. However, questions about whether local knowledge was appropriate to the needs of the technology or the market certainly had to be asked. Responses would point to training needs in the context of fruitful technology use and to improve management effectiveness.

Case of a Closed-system (process) Technology

In this case, a process technology for producing a chemical illustrates the exercises necessary for evaluating the appropriateness of a closed-architecture technology. Because the process is based on knowhow, most of its features will not be in the public domain. Indeed, they may be confidential and will generally only become accessible to the entrepreneur when he enters into a technology licence contract. Product literature or oral information from salespersons is, of course, available for promoting and marketing the technology's product and facilitating its application. It is assumed that the would-be entrepreneur in a developing country has been presented with a single offer of technology and that the offer is not associated with offers of equity or other forms of participation.

The example involves the manufacture of a branded, high tech protective coating for exterior surfaces of all kinds (wood, metal etc). According to a technical brochure enclosed with the marketed product, the coating develops on polymerization of the constituents present in the coating solution. Te polymerizing substances are said to be acrylic esters, with no further qualification. The brochure states that polymerization requires the addition of a mixture of catalysts and other materials packed in a separate container but sold with the ester product. If the process is patented it will be fairly easy to get a good understanding of it because patents generally do four things: (a) disclose "prior art", i.e. how coatings belonging to the patented product group have in general been made; (b) present claims of novelty for the patented product/process; (c) outline the methods available of obtaining the product; and (d) state the preferred mode of making the product. A patent makes the technology more open-architecture.

However, even though all details of the process may be disclosed in the patent, they usually relate to production at the test-tube level or, only details of the critical segment are provided. A capable engineering firm might be able to scale up the process to commercial dimensions, but that would not yield crucial operating knowledge. The questions of how to make the product most economically and with the best specifications would remain open. Operating information, often referred to as know-how or show how, is held in confidence. "Look-see" arrangements, feasible with open-architecture technologies, would in this case reveal very little. Understanding process technology requires knowledge of operating conditions in all segments of the process, not just the patented segment. However, prior disclosure agreements can often be concluded to obtain such information, since the potential for misuse is minimized by the protection already available through the patent.

Contracts for the prior disclosure of process information for technologies that are wholly know-how based are often possible in industrialized countries for a fee. In developing countries, the legal framework may not sufficiently protect an information supplier in terms of ownership rights and wrongful use of process information. To a great extent, in closed-system technologies, a technology analyst has to approach evaluation obliquely and indirectly. The check-list of queries for analyzing closed-system technologies will be broader in scope but poorer in detail than the check-list for open-architecture technology because there is little information in the public domain to use in framing questions. In this case there is no patent in the host country.

It contains queries that a technology proprietor should be able to answer with little fear of violating proprietary information. A search of technical literature in the coatings field may also reveal answers to some of the questions. Some queries raised in the context of the second case that are of a product of higher complexity may also be applicable here. A few kinds of information may not be available during the early stages of technology exploration; these have been marked with an asterisk.

Risk in Technology Transport

The focus thus far has been on the suitability of a technology for use in a developing country environment. That assessment is generally separate from assessing the risks related to the technical performance of a technology that may otherwise be eminently suitable. Technology-associated risks are always present, and their consequences vary insignificance. Some risks may be small, that is, rectifiable at a low cost or with little effort. Other risks may be more difficult to correct (if, for instance, a plant is located near a mine for a raw material but the quality turns out to be poor and raw material must be bought from a distance) but still allow a reasonable profitability. Still others are large enough to cause a venture to be abandoned, e.g. risk of the emission of toxicants forced the closing of some plants in the United States after the Bhopal disaster.

Technology-associated risks arise in several areas. Some key areas have already been referred to: the workability of scaled-down versions of technology and the adaptability of technology to raw materials or utilities with which the technology owner is unfamiliar. These risks are present in all cases of technology transfer. The ultimate user of a technology bears many other risks: incorrect choice of product, insufficient market size, misjudging the market segmentation or product positioning, poor location of production plant, underestimation of investment and so forth. Appraising business risks, which may be greater than technology-associated risks, is peripheral to these analyses. Some risks cannot be covered t all, others may be covered by carefully written contractual provisions, and some may be shared or minimized by involving the technology supplier in the market-place (joint ventures, product-sharing etc). Some risks cannot be controlled, assessed or appraised by either the technology supplier or the recipient. These are accepted by both parties as being uncertainties outside the knowledge or control of the negotiating parties; an example would be an impurity in a raw material.

Risks in Process Industry Technology

The technological risks of closed-system technologies are generally greater than those of open-architecture technologies, or widely used technologies. In the product assembly industries, for example, there are many or few sequential steps in the manufacture of a product, only some of which may result in serious economic risk if improperly assessed. Project-phasing, testing critical equipment prior to shipment and obtaining warranties of the replacement of defective equipment are all risk-reducing measures that may provide early assurances of workability. In the process industries, however, output results from an intricate networking of the constituent elements, all of which must be present and working simultaneously to achieve project objectives. Thus, cement, sugar or paper plants cannot be phased in, nor can any reasonable test be made of an individual piece of equipment without feeding it material from another process unit. An unexplored deficiency in raw material, wrongful use of a construction material, or incorrect configuration in a reaction area can jeopardize an entire project. Another form of risk may lie in all ill-conceived mating of technology supplier and plant engineering/ construction firms.

Another feature of process industries is that the risk characteristics, and the points at which the risks are most significant, are often specific to the industry involved. Thus, in the manufacture of cold-rolled steel, the mechanical properties of the steel and the thickness tolerances obtained may be more critical, and thus a greater risk factor, than the steel's physical or chemical properties or even the output volumes. In the pharmaceuticals industry, a technology may be chosen because it presents the least risk with respect to product purity, shelf-life and clinical performance (e.g. low dosage, few contraindications) rather than for reasons such as yield on raw materials or output stability of the manufacturing process. In the chemical industries, risk exposure may lie in performance parameters such as product yield on raw materials or catalyst stability. Analysis of risk thus involves the identification and analysis of what may be called hidden factors in the industry or technology involved.

Some risks have direct financial implications while other risks, such as public safety aspects of the technology, cannot be measured in these terms. Financial risks may be minimized or shared through mechanisms such s simulating the process at the laboratory level, building a pilot plant, obtaining assurances through process guarantees and warranties, creating a joint venture or building a turnkey plant. In general, risk is minimized when technologies are licensed-in at the mature phase in their life cycle and when the output volume I not too different from that in similar plants. A technology's maturity is indicated by the frequency with which it is being licensed. Industry journals often provide this information about major technologies; alternatively, a licensor may be asked to provide a list of licensees and the dates on which plants constructed under the licences came on stream.

Potential for Environmental Damage

In open-architecture technologies thee is some opportunity for the would-be acquirer to assess its potential to cause environmental damage. In closed-system technologies, the opportunity for such assessment can be quite limited. Disclosure agreements may become mandatory if the technology is suspected of creating an adverse impact in any of these areas; alternatively, affidavits or warranties, at the technology selection stage, may be required. The forms of legal protection that may be available are outside the purview of this module.

At the same time, many open-architecture technologies in which developing countries are interested (e.g. cement, paper or metals) are the developed world's smokestack industries and they can pose great environmental and ecological risks, which must be abated by sophisticated technology. This, however, may turn a previously open-architecture technology into a package with a closed-system component that needs to be analysed as discussed above.

Quantitative Approaches to Assessing Technology Appropriateness and Risk

The availability of more than one technological option has many advantages. A plurality of options, as noted earlier, avoids a Hobson's choice (take it or leave it). It provides alternative routes to manufacturing a product, one of which may be most appropriate for the host country. In many industries there are, at any given time, equivalent technologies competing with each other. The most frequently used methods for selecting one technology from a set of options are those of financial analysis (economic reward). Many kinds of analysis are available, ranging from a simple return on investment analysis to a more complex analysis of internal rate of return. These methods take into account inputs and outputs in terms of costs and prices, the life of the project, time related flows of funds, discount rates of money, inflation and several other factors. They do not, however, weigh technological factors directly.

Financial analysis may weigh some quantitative impacts of appropriateness and risk, but they do not weigh many important qualitative factors bearing on the acceptability of a technology in a particular physical environment and ecosystem. Nevertheless, the economic reward is a fundamental criterion in any analysis of alternative technologies and constitutes a key test of acceptability. Objective methodology must attempt to achieve a better balance between measuring the positive aspects of economic benefit and the negative aspects of inappropriateness and risk. In this module, a variation of the simple return-in-investment (ROI) method, the comparative costing method, is used to determine economic reward.

The exercise is carried out not so much to demonstrate the methodology, which is well-established, but to compare its rating of alternative technologies to the ratings of two other methods: the parameter ranking method and the points system method. Because different aspects of a technology are evaluated using the different methods, it is desirable to use all three to determine the most appropriate technology of those on offer, using information from the suppliers. The recipient must choose a technology with a low-risk profile while trying to obtain maximum possible insurance against risks being accepted, including the risk of selecting an inappropriate technology. Process disclosure agreements, process guarantees and warranties, joint-venture arrangements, shared production and subcontracted manufacturing modes are among the alternatives that may be available.

By that time, the potential technology recipient will have researched independently, or with the help of professional consultants, the technologies in use and through this process, eliminated some alternatives on grounds such as raw materials availability or minimum required plant size. Reaching step G also indicates a technology evaluator has short-listed technologies taking into account the "stickiness" factor, which associates the perspectives of the technology developer with the manner in which a technology will be used.

The Comparative Costing Method

Primary data, which might have been provided in many kinds of units, have been reduced to currency units. While profit before tax (PBT) and PBT/fixed investment have been used as economic comparators, other comparators might also be applied. Technologies may, however, be compared without taking into account financial parameters such as overhead, which vary little with technology, this is known as comparative costing.

If aspects of a technology such as position in the life cycle, impacts on the ecosystem, public hazards and consumer safety are equally favourable for all the technology being compared, the above method would be quite appropriate for industrialized countries, because accessibility to resources is not restricted and market cots (factor prices) are the determining criteria. In developing countries, however, other factors need to be considered. For example, a constraint on foreign exchange might encourage selecting a technology that uses a maximum of indigenous materials (e.g. Capital goods or raw materials); likewise, constraints on natural resources might orient selections to those technologies in which, for example, (hydro)electric power could be substituted for petroleum-based fuels. In these circumstances, the selector may be willing to trade off higher cost and less economic efficiency for minimizing the use of scarce resources.

Again, the disadvantage of the comparative costing method is that it does not provide a mechanism to take into account qualitative factors. The ranking and point systems methods make attempts in this direction.

Ranking Methods

Five criteria are established:

- Fixed investment in national currency to be optimized
- Fuel gas usage to be conserved
- Costs of imported raw materials to be conserved

- Electric power usage to be conserved
- Need for skilled labour to be minimized

Unweighted Ranking

In the first and simplest of these methods, technologies are awarded proficiency marks that are, ranked; with the highest number assigned to the technology most proficient in the use of each parameter, e.g. Maximizes national investment inputs, minimizes the use of natural gas. Likewise, technology E is most proficient in the country's most proficient in the country's use of investment inputs and poor in the use of its raw materials and in conserving fuel gas. While adding proficiency marks might be a useful exercise, it offers little support to realistic analysis because it assigns the same weight to all scarcity factors. It may, however, be a useful tool for comparing investment sites within a country for a particular technology rather than for selecting one of a set of competing technologies.

The technology with the highest weighted cost, that is, the technology that uses scarce resources mot efficiently, is, of course, to be preferred. In this example, technologies D and E are particularly proficient, and when overall cost parameters and the impact on scarce resources are considered, technology D would be preferred. However, selecting it would reduce the economic advantages obtainable by selecting technologies C and A. This then, is the trade off the selector must be able to accept if the priorities (weightings) are significant and are to prevail. Ranking methods are useful when critical parameters can be quantified on a rational basis and weights can be assigned. However, they are relatively inefficient when there are a large number of qualitative factors.

The Points System Method

The points systems method takes into account the qualitative factors cited in check-lists 1 and 2 (e.g. operational, public safety) that cannot be quantified or weighted.

However, it, like the ranking methods, involves problems of subjectivity. These problems will be dealt with after describing the points systems method. The following steps are involved:

- Key evaluation parameters are listed and evaluation criteria are clearly defined.
- The parameter the selector considers most significant – the reference parameter - is assigned a weight of 100.
- The weights of the other parameters are assigned by the selector considering their importance compared to the reference parameter (they will, by definition, be less than 100). This gives rise to point system scale.
- One of the candidate technologies is taken as the reference technology. It can be any one of the technologies being considered.
- For this reference technology, and using the points system scale, the selector attempts to establish a point score by assigning the maximum number of point if it is less favourable. This establishes a vertical scoring component.
- With the reference technology thus scored all other candidate technologies are compared to it, parameter by parameter, and scored. Some technologies may get a higher score than the reference technology. This is the horizontal scoring component of the methodology.

Assessment in the Dual-bid Method

Many developing country agencies use what may be termed the dual-bid, or double-envelope, method for selecting technology. In this method, a short-listed group of licensors makes two-part sealed bids. The first – the technical bid – details the offering in terms of technology proficiency factors, and the second – the cash or commercial bid – identifies the fixed investment and technology costs for the technology package. Bids are formulated using a questionnaire prepared by the potential buyer, using consultants if necessary. Evaluators on the buyer side then further short-list the technologies from the economic and technical proficiency perspectives, taking available resources into account.

The technologies are reviewed separately, by financial experts and decision makers, from the commercial and business points of view. When this type of bidding process is used, the ranking methods and the points system method are particularly relevant for analyzing the technical bid, leaving the comparative costing methods until the last. To carry out the assessments suggested in this module, a significant amount of data and information about technologies is required from the owners. Generally, this becomes possible after confidence is established that the analyst is serious and that one of a proffered set of technologies will finally be selected. These assessment methods use information technology owners are usually willing to divulge to technology evaluators.

Testing Subjectivity

As pointed out earlier, and as evident from the ranking methods and the point system method, there is likely to be a substantial degree of subjectivity in an analysis, both in the selection of the parameters and in the scoring. Fortunately, several statistical methods are available to test the degree of subjectivity in analysis. They can be used to assess the selection of parameters, scoring or both. Two of the easiest methods are illustrated here. It needs to be pointed out that, to be as objective s possible, those who evaluate the technologies and hose who select the parameters and establish weight s for them, an correspondingly for the points system scale, must be different people. The selection of parameters and their weighting should be done by senior managers or teams experienced in the technical and economic aspects of technology. This would remove one of the several subjective factors inherent in such exercises.

Technological Complexity and Technology Transfer

It has been assumed in the course of the above discussion that if a selected technology meets certain techno-economic criteria it is appropriate and can be transported from the country of its development (and use). Many of these criteria have been outlined: a technology's adaptability to smaller markets, its accommodation scarce resources, its adherence to certain qualitative criteria, its maintainability given the skill levels of the new habitat, etc. They are important, but sometimes insufficient, conditions for successful transfer. A key criterion that must also be assessed is the workability of transported technology in the context of the technological complexity in both the sourcing and host countries. Technological complexity relates to the manner in which and the extent to which technology is used to yield output and diversity of modern, goods and services, and to carry out the tasks of industrial management and organization, as well as on the means adopted for its development, propagation, permeation and protection.

In are beneficial aspects technological complexity ultimately manifests itself in the form of products and services that reduce drudgery in carrying out everyday work provide greater comfort and convenience, afford more time for the pursuit of leisure activities, and so forth. Hence, a high degree of technological complexity typically reflects a high quality of life. Experience shows that unless certain externalities are similar in the two environments, many of the micro-economic benefits exhibited by an individual technology in the sourcing country or environment will not be realized in the host country environment. In other words, unlike water, technology does not flow well from a high level to a low level environment.

The Technological Infrastructure

The technological complexity of an economy evolves gradually. At its high end, the substantive components of technological complexity are manifest in external facets such as replacing or supplementing natural products with synthetic substitutes, high manufacturing and processing speeds consistent with high levels of product quality, progressive integration and assembly of parts, miniaturization and the increasing substitution of machines and systems for human skills. With respect to the last item, one may cite the substitution of machines for muscle-embodied labour, of automation and robotization for skilled labour, of computers for clerical and many categories of supervisory personnel, and of artificial intelligence and neural systems for middle-level managers.

Technological complexity is further demonstrated by the use of technology to network goods and services. Goods are ordered, paid for, inventoried, and employed in efficient systems through the use of information technology, with a few people controlling the movement of large volumes of goods through a complex transport system. There are also many systems to deliver services: human hierarchies within an enterprise perform certain functions and specialized professional agencies outside the enterprise to perform others. The systems by which technology is produced, employed and licensed contribute to technological complexity. Legal instruments such as patents, trade marks and trade secrets legislation enable relatively easy access to technology. Technological complexity both reflects and is enhanced by the presence of a well-developed technological infrastructure.

A high-quality industrial infrastructure supports a large number of business transactions per unit of time over distances by a variety of means, such as computer, fax, telephone, video-conferencing, person-to-person interactions. Indeed, the relative levels of two industrial infrastructures can be measured by the number of transactions they can support per capita: (number of two-party transactions per unit of time) x (the sum of the distances separating the transacting parties)/ population.

The Transportability of Technology

The transportability of a technology, that is, the ability to relocate it to another environment, depends on the technological complexity of the national (industrial-economic) environment and of the particular industry in which it will be imbedded. Of these, the complexity at the national level is perhaps secondary. Thus it is possible for particular groups of advanced technologies to work adequately and efficiently within an industry having a high technological complexity internally even though, the country lags well behind those countries from which the technologies have been sourced. The effective performance of computer based technologies in Taiwan Province of China and the Republic of Korea is an example. Since computer technology is one focus development in these countries industries employing these particular technologies have thrived and been efficient despite their need for extreme levels of miniaturization and processing speeds (which are characteristic of many industries in the technologically advanced countries).

The same is true for the incorporation of advanced textile industry technologies in India, which call for high levels of vertical and horizontal integration of industry systems, a multifibre processing capability, and large volumes of production. That having been said, it remains true that, technology flows most effectively from one point to another when the levels of technological complexity are nearly the same. The degrees to which human skills can be replaced by machines and systems, the extent to which goods and services are networked and the transaction capability of the technological infrastructure are three external features that determine the transportability of technology from one environment to another.

It might be asked. Does the reverse situation hold? Does a technology efficient in a place of lower technological complexity perform satisfactorily when transported to a place at a higher level? Specific examples to illustrate the point are difficult to find. One, however, is the superior performance (i.e. higher output per main-hour of the work) of software people (a non-material form of technology transfer) when they move from a developing country to a developed one.

Corporative Significance

If we look at technological complexity in the United States and set it at an arbitrary level of 100, then transfer of technology, at any point on the technology life cycle, to Western Europe and Japan may be expected to be as effective and rewarding as transfers within the United States. If, on the other hand, the same technologies wee to be transferred to a newly industrializing country with a level of technological complexity of 50-60, then the effectiveness of transfer might prove poorer. Even so, it might be more effective than transfers to; for example, a developing country in Africa with a level of complexity in the 15-20 range, where the technical and economic conditions will be unacceptable and the technologies will fail to perform.

On the other hand, a technology transfer from one developing country to another, with a level of technological complexity, will have less possibility of distortion through reconfiguration than a transfer from a country with a markedly higher level of technological complexity. That is, if a technology is transferred from an environment with a technology complexity factor of 40 (relative to the United States) to one that has a technology complexity factor of 30 (relative to the United States) then it will travel well, particularly if adapted to national endowments.

However, technologies that have existed for a long time and are in the declining phase of their technology life cycle may be transportable to locations with a substantially lower level of technological complexity without significant potential of distortion through reconfiguration. This is true because, first, the technology in the declining phase was developed when the environment was at a lower level of technological complexity without significant potential of distortion through reconfiguration. This is true because, first, the technology in the declining phase was developed when the environment was at a lower level of complexity than at the time of its transfer, and secondly the technology has by now become a technique (that is, a specialized skill) and thus carries little risk of inadequate performance. Much of the technology flowing between advanced and developing countries may be characterized as technique which is why it works well in new habitats. These transactions can be said to involve technical services rather than true technology.

One of the paradoxes of modern-day economic reality is that developing country with a low level of technological complexity require certain high-level technologies, such as power and telecommunications systems, or systems to exploit natural resources for exports. Because such technologies need not be scaled-down or modified to suit the factor endowments of the recipient country, they transfer well, particularly when the transfers are made on a turnkey basis. At the same time, they are not readily absorbed by hot country technicians and managers and will continue to require external maintenance support for optimum perfor-mance. Thus, consciously acknowledging the importance of the relative technological complexities of both sourcing and receiving countries permits one to assess the transportability of a technology. In most cases, a qualitative determination suffices; the difference in technological complexities is too great to successfully transfer technology, or the complexities are of a comparable level, or the penalties of the difference can (or cannot) be absorbed at a moderate cost.

An assessment Tool

Sometimes, however, an objective method of assessing the relative levels of technological complexity between the technology-source and the technology-recipient countries is called for. One way of doing this is to take poll o experts and analyse its findings using the Delphi Principle. The method suggested here is analogous to, and derived from the points system method. It is a simple way of looking at technological complexity to support decision making for technology transfer.

Such as analysis is easier and the results are clearer, when evaluating a technology entering a particular industry. Here, however, it is the economic industrial technological complexity of the two countries that are being compared.

The basic methodology comprises the following steps:

- List the external features needed to support the successful operation of the technology in the source country (country A).
- For the country that will receive the technology (country (B), give a rating of 100 to each feature. If the feature is absent, give a score of zero.
- Rate each feature in country A. The score will generally, but not always, be higher than 100. For instance, if the quality of telecommunications was the feature being assessed and its score was 100 for country B, it could well be that the score for that feature in country A would be 400. Likewise, for transportation flexibility, the scores could be 100 (for country B) and 250 (for country A). On the other hand, in terms of accessibility to unskilled labour, the score for country A might be 20 compared to 100 for country B and perhaps zero for access to certain raw materials. When the country B score is zero, prorate the country A score looking at the scores you have given to other features of country A.
- Total the points for each transacting country.
- Set the technology complexity factor 100 for the country acting as the technology source (country A).
- Obtain the proportional comparative factor of technology complexity for the country receiving the technology (it will generally be below 100).

- Assess the impact of this factor on the transportability of the technology between the two countries.

The important external features that can affect the performance of a technology may be listed as follows:

Industrial system:

- Degree of industrialization, i.e. number of industries, by type, in the country
- Degree of horizontal and vertical integration
- Geographic dispersion/concentration of industries
- Interdependence of products and services and the degree of networking of products and services

Technology system:

- Intensity of replacement of labour of various skill levels by machines, automation and computerization
- Complexity and depth of the technology information system

Status of intellectual property rights:

- Knowhow
- Patents
- Trade marks and designs
- Copyright protection (for software, etc)

Marketing system:

- Size of markets
- Complexity of product mix
- Degree of competitiveness within industries
- Competition from imported products and services
- Technology of the distribution system
- Technical servicing capabilities
- Manpower system
- Availability of unskilled and skilled labour
- Availability and cadres of supervisory and managerial personnel

Institutional structures:

- Technical schools
- University and corporate R&D centers
- Design and engineering firms
- Construction and erection forms
- Product and technology consultancy organizations
- Role of national government in institutional structures

Infrastructure:

- Accessibility of raw materials and utilities
- Transport systems
- Telecommunications

Basis of Calculating Equivalent Technology Cost

Using the concept of present value, it is possible to reduce various expressions of time-related technology fees to a common, comparative basis. Each future payment is reduced to its present value by discounting it at a discounting rate, which may vary between countries. That is, at a 10 per cent discount rate, $1.00 received a year from now is equal to $0.9091 today (its present value) Where a comparison being made between technologies relative to their respective ascendancy, the application of a 10 per cent will not unduly distort results. The UNIDO publication "Guidelines for the Evaluation of Transfer of Technology Agreements," DTT series, No. 12 (1979), provides more background to this methodology, which may be used to reduce to comparative figures.

With a three-year, $1.1 million grant from the National Science Foundation (NSF), leading research groups plan to create a standard for evaluating how desktop, laptop, and handheld computers are currently used in mathematics and science education—with the ultimate goal of advancing the idea of ubiquitous (one-to-one) computing. "What are teachers doing with laptops and handhelds in math and science classes? It's a very simple question, but no one knows," said Andy Zucker, principal investigator for the Ubiquitous Computing Evaluation Consortium and associate director at SRI International's Center for Education Policy. As more schools integrate one-to-one computing into the classroom, it becomes increasingly important to determine how the devices are being used, how ubiquitous computing changes the learning experience, and how teacher's best can integrate available technology into their curricula, Zucker said.

Creating a standard framework for research enables evaluators, policy makers, and educators to compare and tally the outcome of various studies more easily and accurately. "By pooling our resources and sharing research findings with the six other institutions in the consortium, we'll be able to develop the comprehensive framework and the knowledge that] policy makers need to understand and guide the widespread adoption of computers in math and science classrooms," Zucker said. "We will be able to make quicker headway." Within a few months, the consortium will post the first draft of its framework for evaluating ubiquitous computing on the project's web site and invite the public to comment on it.

SRI International, an independent not-for-profit research and development organization, is coordinating the effort. So far, the consortium includes several well-known institutions in the ed-tech field, including the University of Virginia, the University of Minnesota, the Metiri Group, Kent State University, Rockman et al, and the Education Development Center's Center for Children and Technology. In addition to collaborating on a national level, each organization is working separately to evaluate a number of ubiquitous computing initiatives in several states. "Each of us will be doing one-to-one computing evaluations at different sites. That means there will be at least seven studies underway," Zucker said. These studies are not funded under the NSF grant, but will adhere to the guidelines established by the consortium, and their findings will be published on the project's web site.

In addition, the site will feature other tools and resources designed to promote ubiquitous computing in schools. The consortium is developing a teacher survey for school officials to use. "Once we have a teacher survey we think is good, we will put it up on the web site and anyone will be free to use it," Zucker said. The Metiri Group is conducting a policy study to determine what policy makers need to know about ubiquitous computing. The group will interview dozens of policy makers, including staff from governor's offices, school superintendents, and others.

Traditional Paradigm and its Problems

It's easy to describe the ideal way to evaluate a grant to use technology to improve education:

1. First, figure out the desired impact on people's lives, what will matter to them (and to the funder, perhaps).
2. Then figure out a way to measure that attribute in people.
3. When the new software or approach is implemented, do a pretest of that attribute and then measure again after the grant has had its impact on them. At the same time, do a similar pair of tests on comparable people going through traditional education of this sort. Discover, if you're lucky, that the innovative program has resulted in more learning than the traditional method.
4. Turn in this hard data to the funder as part of your final report on your grant and disseminate it in the wider world.

That paradigm sounds simple (although not easy) to do. Unfortunately, a number of fundamental problems often make that sort of evaluation impossible or inadequate, and potentially misleading. In this essay, we'll consider a few of those difficulties with the paradigm and what sorts of approaches can complement or replace it.

1. *The Project and its Grant-Funded Evaluation are usually over before much of the Project's Impact on People's Lives Has Begun*

   Ordinarily, development projects have a chance to make a trial run or two, but the kinds of ultimate impacts that funders care about may well not have happened by the time evaluators must stop collecting data and start writing reports. Thus, the report usually dwells on problems of implementation and "feel good" stories about process.

   Response: Extend the grant well beyond the end of the rest of the project. Or fund another project later on whose chief purpose is evaluation.

2. *The Paradigm implies that "Traditional" Education is Uniform and Well Understood*

   "Traditional education" as a label covers even more ground than the label "innovative." Many evaluations are flawed in that they provide virtually no description of what was done in the control (traditional) group. Response: Be just as searching in the evaluation of the "traditional" as in the evaluation of the "innovation" and just as descriptive in the report.

3. *The Paradigm implies that we Understand What this Innovation is and What it's for, in Advance*

   The paradigm implies foreknowledge because it requires the development of a pretest that students take before using the innovation (or being used by it) and a parallel post-test. That's not possible if one isn't clear about what is to be taught. What's so difficult about that? A first problem with this image is the problem of the moving target: the use of technology usually is associated with a change in the aims of an assignment, course, or course of study. Statistics provides just one example of this problem.

   A couple of decades ago, it was normal to teach statistics to students who used only paper and pencil to do assignments: that places certain limits on the techniques that could be taught. Today's statistics course (taught to students using advanced calculators or other computers) not only teaches new statistical techniques that require such machines; it also teaches new skills of thinking (e.g., strategies for investigating data by analyzing images created with the data). To evaluate the impact of computing on learning over the last couple of decades, one would have needed to anticipate the new techniques and skills and somehow incorporated them or their analogue in tests given years earlier. Without that, one is faced with a meaningless comparison — students scored an average of 80% on their exams two decades ago and still average 80% (on fundamentally different exams). Are they learning "more"? The "same"? "Less"? The question has no meaning.

Response: There is no completely satisfactory response. One might:

1. Evaluate whether the conditions of learning have improved (e.g., time on task is usually correlated with improved learning outcomes, so has time on task improved?);
2. Evaluate the tests themselves (if students averaged, say, a "B" five years ago and average a "B" today, do external evaluators believe that this year's tests and assignments are more up to date and perhaps tougher than five years ago? Or have they been "dumped down?").

A second problem about assuming an understanding of the innovation is that innovative projects evolve, often relatively rapidly. Innovations change as the implications of the underlying ideas emerge. They change as the capabilities of new hardware and software become well understood. They change as external decision makers perceive and react to the innovation.

Response: It's always a good idea to periodically step back and ask, "What is this initiative? What is it for? Where is it going?" The answers may well have changed in surprising ways over a period of months. A project whose chief objective a year ago was to help students understand the sociology of organization may well have acquired a new goal of instilling multicultural insight in the last twelve months.

A third problem relating to the identity of the innovation might be called "rapture of the technology." The evaluator erroneously assumes that the technology is the innovation. A project a decade ago provides a dramatic illustration of this mistake. In the mid 1980s, Brown University began developing Intermedia. (If you don't happen to remember it, visualize today's World Wide Web — in fact something better than today's Web in the ease of creating links in a shared body of text and graphics — but existing only on a local area network.) Brown's project was ambitious: to develop Intermedia and test it by letting faculty develop course materials in it and then teach in it. Ultimately, two professors were chosen to participate in the experiment as development of Intermedia got underway. One of the two was George Landow, a full professor in the Department of English.

The project had an unusually large fraction of its grant devoted to evaluation. Step 1 was for a team of anthropologists to observe Landow's literature course in its primal state: pre-Intermedia. Step 2 was for them to observe his whole course again, a year later, when he was using the prototype Intermedia system. In the interim, Landow advised on Intermedia development and crafted his course materials — a web of essays and graphics on Dickens, literature, and the Victorian period. There was a slip between Step 1 and Step 2, unfortunately: Intermedia fell behind schedule and was not ready for use in time.

The evaluators were scheduled to observe Landow, however, so they did, even though he was still teaching without technology. They were surprised to observe that he had become a much better teacher: the experience of developing courseware had changed Landow — he used more graphics, he had students creating pictures of the linkages between concepts, and he put more emphasis on student participation in the seminar. (When they observed Landow again, a year later when he did have Intermedia, the course had improved still further.)

The point here is that Intermedia itself was only part of the innovation: the full innovation, in Landow's hands at least, was a more learner-centered, more visual, more associative approach to the teaching of this literature course. Intermedia was a powerful tool that helped him implement this approach. Response: Be systemic in defining the boundaries of the innovation. It might include not only software but staff (and their skills), changes in infrastructure, changes in the "market," and so on.

4. *The Paradigm implies that the Project's Objectives are the same for Everyone*

There are two ways to look at almost any educational program, both valid:

- As an effort to impact all learners in the same way (though some learners will almost always achieve more in that direction than others); and
- As a resource that different learners will use differently with qualitatively different and somewhat unpredictable consequences.

The first perspective can be called "uniform impact" because it assumes that the educator is trying to shape learning and to do so in the same way for all learners. For example, when an English faculty member grades students on grammar, a uniform impact standard is probably being applied.

The second perspective can be called "unique uses" because it assumes that each learner actively interprets and makes use of the resource. Because human beings are unique, because of accidents, and because the teacher and other students will react to this learner, the educational outcomes will be particular to that learner. What matters is how good the learning outcome is (within a broad range of possibilities) rather than whether it matches a behavioral standard being applied to all students. For example, when an English faculty member grades student essays and gives two students B's for completely different reasons, a unique uses perspective is probably being applied.

Each of these two perspectives implies a different approach to evaluation. The paradigm described at the start of this essay embodies a uniform impact approach: state the goal for all students in advance, and then test their learning. The unique uses perspective on that same innovation can't use that approach to evaluation. The same test can't be used for all students because one is searching for learning outcomes that may well be qualitatively different from one student to the next. Instead, a unique uses evaluation begins with a broad search for important outcomes, good and bad, and then assesses each case one at a time. The English faculty member assigns an essay topic, "What I did on my Summer Vacation," and then grades the essays one at a time.

The accreditor asks certain broad questions, reads the self-studies, and then visits the campuses to examine how well each one is achieving its own goals. First the cases are assessed individually (What were the most important outcomes for this student? How plausible is it that what I'm seeing is indeed an outcome of the program we're evaluating?). Then the evaluator considers what they imply, collectively, about the innovation.

6

Science, Scientific Method, and the Authority of Experts

Scientific method refers to bodies of techniques for investigating phenomena, acquiring new knowledge, or correcting and integrating previous knowledge. To be termed scientific, a method of inquiry must be based on gathering observable, empirical and measurable evidence subject to specific principles of reasoning. A scientific method consists of the collection of data through observation and experimentation, and the formulation and testing of hypotheses. Although procedures vary from one field of inquiry to another, identifiable features distinguish scientific inquiry from other methodologies of knowledge. Scientific researchers propose hypotheses as explanations of phenomena, and design experimental studies to test these hypotheses.

These steps must be repeatable in order to dependably predict any future results. Theories that encompass wider domains of inquiry may bind many hypotheses together in a coherent structure. This in turn may help form new hypotheses or place groups of hypotheses into context. Among other facets shared by the various fields of inquiry is the conviction that the process be objective to reduce a biased interpretation of the results. Another basic expectation is to document, archive and share all data and methodology so they are available for careful scrutiny by other scientists, thereby allowing other researchers the opportunity to verify results by attempting to reproduce them. This practice, called full disclosure, also allows statistical measures of the reliability of these data to be established.

Scientific Method

Since Ibn al-Haytham (Alhazen, 965–1039), one of the key figures in developing scientific method, the emphasis has been on seeking truth:

"Truth is sought for its own sake. And those who are engaged upon the quest for anything for its own sake are not interested in other things. Finding the truth is difficult, and the road to it is rough."

"How does light travel through transparent bodies? Light travels through transparent bodies in straight lines only.... We have explained this exhaustively in our *Book of Optics*. But let us now mention something to prove this convincingly: the fact that light travels in straight lines is clearly observed in the lights which enter into dark rooms through holes.... The entering light will be clearly observable in the dust which fills the air."

The conjecture that "light travels through transparent bodies in straight lines only" was corroborated by Alhazen only after years of effort. His demonstration of the conjecture was to place a straight stick or a taut thread next to the light beam, to prove that light travels in a straight line. Scientific methodology has been practiced in some form for at least one thousand years. There are difficulties in a formulaic statement of method, however. As William Whewell (1794–1866) noted in his *History of Inductive Science* (1837) and in *Philosophy of Inductive Science* (1840), "invention, sagacity, genius" are required at every step in scientific method. It is not enough to base scientific method on experience alone; multiple steps are needed in scientific method, ranging from our experience to our imagination, back and forth.

In the twentieth century, a hypothetical-deductive model for scientific method was formulated:

1. Use your experience-Consider the problem and try to make sense of it. Look for previous explanations. If this is a new problem to you, then move to step 2.
2. Form a conjecture-When nothing else is yet known, try to state an explanation, to someone else, or to your notebook.
3. Deduce a prediction from that explanation-If you assume 2 is true, what consequences follow?
4. Test-Look for the opposite of each consequence in order to disprove 2. It is a logical error to seek 3 directly as proof of 2. This error is called *affirming the consequent.*

This model underlies the scientific revolution. One thousand years ago, Alhazen demonstrated the importance of steps *1* and *4*. Galileo (1638) also showed the importance of step *4* (also called Experiment) in *Two New Sciences*. One possible sequence in this model would be *1, 2, 3, and 4*. If the outcome of *4* holds, and *3* is not yet disproven, you may continue with *3, 4, 1,* and so forth; but if the outcome of *4* shows *3* to be false, you will have go back to *2* and try to invent a *new 2*, deduce a *new 3*, look for *4*, and so forth.

Note that this method can never absolutely verify (prove the truth of) 2. It can only falsify 2. (This is what Einstein meant when he said "No amount of experimentation can ever prove me right; a single experiment can prove me wrong.")

In the twentieth century, Ludwik Fleck (1896–1961) and others found that we need to consider our experiences more carefully, because our experience may be biased, and that we need to be more exact when describing our experiences. These considerations are discussed below.

DNA Example

The Keystones of Science project, sponsored by the journal *Science*, has selected a number of scientific articles from that journal and annotated them, illustrating how different parts of each article embody scientific method. Here is an annotated example of this scientific method example titled *Microbial Genes in the Human Genome: Lateral Transfer or Gene Loss?*

Each element of scientific method is illustrated below by an example from the discovery of the structure of DNA:

- DNA-characterizations-In this case, although the significance of the gene had been established, the mechanism was unclear to anyone, as of 1950.
- DNA-hypotheses-Crick and Watson hypothesized that the gene had a physical basis - it was helical.
- DNA-predictions-From earlier work on tobacco mosaic virus, Watson was aware of the significance of Crick's formulation of the transform of a helix. Thus he was primed for the significance of the X-shape in photo 51.
- DNA-experiments-Watson sees photo 51.

The examples are continued in "Evaluations and iterations" with *DNA-iterations*.

Truth and Belief

Belief can alter observations; those with a particular belief will often see things as reinforcing their belief, even if they do not. Needham's *Science and Civilization in China* uses the 'flying horse' image as an example of observation: in it, a horse's legs are depicted as splayed, when the stop-action picture by Eadweard Muybridge shows otherwise. Note that at the moment that no hoof is touching the ground, the horse's legs are gathered together and are not splayed, but for when a horse is jumping.

Earlier paintings depict the incorrect flying horse observation. This demonstrates Ludwik Fleck's caution that people observe what they expect to observe, until shown otherwise; our beliefs will affect our observations (and therefore our subsequent actions). The purpose of the scientific method is to test a hypothesis, a proposed explanation about how things are, via repeatable experimental observations which can contradict the hypothesis so as to fight this observer bias.

Elements of Scientific Method

There are many ways of outlining the basic method shared by all fields of scientific inquiry. The following examples are typical classifications of the most important components of the method on which there is wide agreement in the scientific community and among philosophers of science. There are, however, disagreements about some aspects.

The following set of methodological elements and organization of procedures tends to be more characteristic of natural sciences than social sciences. In the social sciences mathematical and statistical methods of verification and hypotheses testing may be less stringent. Nonetheless the cycle of hypothesis, verification and formulation of new hypotheses will resemble the cycle described below.

The essential elements of a scientific method are iterations, recursions, interleaving, and orderings of the following:

- Characterizations (observations, definitions, and measurements of the subject of inquiry).
- Hypotheses (theoretical, hypothetical explanations of observations and measurements of the subject).
- Predictions (reasoning including logical deduction from the hypothesis or theory).
- Experiments (tests of all of the above).

Each element of a scientific method is subject to peer review for possible mistakes. These activities do not describe all that scientists do but apply mostly to experimental sciences (e.g., physics, chemistry). The elements above are often taught in the educational system. Scientific method is not a recipe: it requires intelligence, imagination, and creativity.

It is also an ongoing cycle, constantly developing more useful, accurate and comprehensive models and methods. For example, when Einstein developed the Special and General Theories of Relativity, he did not in any way refute or discount Newton's *Principia*.

On the contrary, if the astronomically large, the vanishingly small, and the extremely fast are reduced out from Einstein's theories — all phenomena that Newton could not have observed — Newton's equations remain. Einstein's theories are expansions and refinements of Newton's theories, and observations that increase our confidence in them also increase our confidence in Newton's approximations to them. A linearized, pragmatic scheme of the four points above is sometimes offered as a guideline for proceeding:

1. Define the question
2. Gather information and resources (observe)
3. Form hypothesis
4. Perform experiment and collect data
5. Analyze data
6. Interpret data and draw conclusions that serve as a starting point for new hypothesis
7. Publish results
8. Retest (frequently done by other scientists)

The iterative cycle inherent in this step-by-step methodology goes from point 3 to 6 back to 3 again. While this schema outlines a typical hypothesis/testing method, it should also be noted that a number of philosophers, historians and sociologists of science (perhaps most notably Paul Feyerabend) claim that such descriptions of scientific method have little relation to the ways science is actually practiced.

The "operational" paradigm combines the concepts of operational definition, instrumentalism, and utility:

The essential elements of a scientific method are operations, observations, models, and a utility function for evaluating models.

- Operation-Some action done to the system being investigated.
- Observation-What happens when the operation is done to the system.
- Model-A fact, hypothesis, theory, or the phenomenon itself at a certain moment.
- Utility Function-A measure of the usefulness of the model to explain, predict, and control, and of the cost of use of it. One of the elements of any scientific utility function is the refutability of the model. Another is its simplicity, on the Principle of Parsimony also known as Occam's Razor.

Characterisations

Scientific method depends upon increasingly more sophisticated characterizations of the subjects of investigation. (The *subjects* can also be called *unsolved problems* or the *unknowns*.) For example, Benjamin Franklin correctly characterized St. Elmo's fire as electrical in nature, but it has taken a long series of experiments and theory to establish this. While seeking the pertinent properties of the subjects, this careful thought may also entail some definitions and observations; the observations often demand careful measurements and/or counting.

- "I am not accustomed to saying anything with certainty after only one or two observations." —Andreas Vesalius (1546)

The systematic, careful collection of measurements or counts of relevant quantities is often the critical difference between pseudo-sciences, such as alchemy, and a science, such as chemistry or biology. Scientific measurements taken are usually tabulated, graphed, or mapped, and statistical manipulations, such as correlation and regression, performed on them. The measurements might be made in a controlled setting, such as a laboratory, or made on more or less inaccessible or unmanipulatable objects such as stars or human populations. The measurements often require specialized scientific instruments such as thermometers, spectroscopes, or voltmeters, and the progress of a scientific field is usually intimately tied to their invention and development.

Uncertainty

Measurements in scientific work are also usually accompanied by estimates of their uncertainty. The uncertainty is often estimated by making repeated measurements of the desired quantity. Uncertainties may also be calculated by consideration of the uncertainties of the individual underlying quantities that are used. Counts of things, such as the number of people in a nation at a particular time, may also have an uncertainty due to limitations of the method used. Counts may only represent a sample of desired quantities, with an uncertainty that depends upon the sampling method used and the number of samples taken.

Measurements demand the use of *operational definitions* of relevant quantities. That is, a scientific quantity is described or defined by how it is measured, as opposed to some more vague, inexact or "idealized" definition. For example, electrical current, measured in amperes, may be operationally defined in terms of the mass of silver deposited in a certain time on an electrode in an electrochemical device that is described in some detail. The operational definition of a thing often relies on comparisons with standards: the operational definition of "mass" ultimately relies on the use of an artifact, such as a certain kilogram of platinum-iridium kept in a laboratory in France.

The scientific definition of a term sometimes differs substantially from its natural language usage. For example, mass and weight overlap in meaning in common discourse, but have distinct meanings in mechanics. Scientific quantities are often characterized by their units of measure which can later be described in terms of conventional physical units when communicating the work. New theories sometimes arise upon realizing that certain terms had not previously been sufficiently clearly defined. For example, Albert Einstein's first paper on relativity begins by defining simultaneity and the means for determining length.

These ideas were skipped over by Isaac Newton with, *"I do not define time, space, place and motion, as being well known to all."* Einstein's paper then demonstrates that they (viz., absolute time and length independent of motion) were approximations. Francis Crick cautions us that when characterizing a subject, however, it can be premature to define something when it remains ill-understood. In Crick's study of consciousness, he actually found it easier to study awareness in the visual system, rather than to study free will, for example. His cautionary example was the gene; the gene was much more poorly understood before Watson and Crick's pioneering discovery of the structure of DNA; it would have been counterproductive to spend much time on the definition of the gene, before them.

Example of Characterisations

DNA-characterisations

The history of the discovery of the structure of DNA is a classic example of the elements of scientific method: in 1950 it was known that genetic inheritance had a mathematical description, starting with the studies of Gregor Mendel. But the mechanism of the gene was unclear. Researchers in Bragg's laboratory at Cambridge University made X-ray diffraction pictures of various molecules, starting with crystals of salt, and proceeding to more complicated substances. Using clues which were painstakingly assembled over the course of decades, beginning with its chemical composition, it was determined that it should be possible to characterize the physical structure of DNA, and the X-ray images would be the vehicle.

The characterization element can require extended and extensive study, even centuries. It took thousands of years of measurements, from the Chaldean, Indian, Persian, Greek, Arabic and European astronomers, to record the motion of planet Earth. Newton was able to condense these measurements into consequences of his laws of motion. But the perihelion of the planet Mercury's orbit exhibits a precession that is not fully explained by Newton's laws of motion. The observed difference for Mercury's precession between Newtonian theory and relativistic theory (approximately 43 arc-seconds per century), was one of the things that occurred to Einstein as a possible early test of his theory of General Relativity.

Hypothesis Development

A hypothesis is a suggested explanation of a phenomenon, or alternately a reasoned proposal suggesting a possible correlation between or among a set of phenomena. Normally hypotheses have the form of a mathematical model. Sometimes, but not always, they can also be formulated as existential statements, stating that some particular instance of the phenomenon being studied has some characteristic and causal explanations, which have the general form of universal statements, stating that every instance of the phenomenon has a particular characteristic.

Scientists are free to use whatever resources they have — their own creativity, ideas from other fields, induction, Bayesian inference, and so on — to imagine possible explanations for a phenomenon under study. Charles Sanders Peirce, borrowing a page from Aristotle (*Prior Analytics*, 2.25) described the incipient stages of inquiry, instigated by the "irritation of doubt" to venture a plausible guess, as *abductive reasoning*. The history of science is filled with stories of scientists claiming a "flash of inspiration", or a hunch, which then motivated them to look for evidence to support or refute their idea. Michael Polanyi made such creativity the centerpiece of his discussion of methodology.

William Glen observes that:

the success of a hypothesis, or its service to science, lies not simply in its perceived "truth", or power to displace, subsume or reduce a predecessor idea, but perhaps more in its ability to stimulate the research that will illuminate ... bald suppositions and areas of vagueness.

In general scientists tend to look for theories that are "elegant" or "beautiful". In contrast to the usual English use of these terms, they here refer to a theory in accordance with the known facts, which is nevertheless relatively simple and easy to handle. Occam's Razor serves as a rule of thumb for making these determinations.

DNA-hypotheses

Linus Pauling proposed that DNA might be a triple helix. Francis Crick and James Watson learned of Pauling's hypothesis, understood from existing data that Pauling was wrong and realized that Pauling would soon realize his mistake. So, the race was on to figure out the correct structure— except that Pauling did not realize at the time that he was in a race!

Predictions from the Hypothesis

Any useful hypothesis will enable predictions, by reasoning including deductive reasoning. It might predict the outcome of an experiment in a laboratory setting or the observation of a phenomenon in nature. The prediction can also be statistical and only talk about probabilities. It is essential that the outcome be currently unknown. Only in this case does the eventuation increase the probability that the hypothesis be true. If the outcome is already known, it's called a consequence and should have already been considered while formulating the hypothesis. If the predictions are not accessible by observation or experience, the hypothesis is not yet useful for the method, and must wait for others who might come afterward, and perhaps rekindle its line of reasoning. For example, a new technology or theory might make the necessary experiments feasible.

DNA-predictions

The hypothesis (by James Watson and Francis Crick among others) that DNA had a helical structure implied the prediction that it would produce an x shaped X-ray diffraction pattern. This followed from the work of Cochran, Crick and Vand (and independently by Stokes) who had provided a mathematical basis for the empirical observation that helical structures produce x shapes. Also in their first paper, Watson and Crick predicted that the double helix structure provided a simple mechanism for DNA replication, writing "It has not escaped our notice that the specific pairing we have postulated immediately suggests a possible copying mechanism for the genetic material".

General Relativity

Einstein's theory of General Relativity makes several specific predictions about the observable structure of space-time; such as a prediction that light bends in a gravitational field and that the amount of bending depends in a precise way on the strength of that gravitational field. Arthur Eddington's observations made during a 1919 solar eclipse supported General Relativity rather than Newtonian gravitation.

Experiments

Once predictions are made, they can be tested by experiments. If test results contradict predictions, then the hypotheses are called into question and explanations may be sought. Sometimes experiments are conducted incorrectly and are at fault. If the results confirm the predictions, then the hypotheses are considered likely to be correct but might still be wrong and are subject to further testing. The experimental control is a technique for dealing with observational error. This technique uses the contrast between multiple samples (or observations) under differing conditions, to see what varies or what the same remains. We vary the conditions for each measurement; to help isolate what has changed. Mill's canons can then help us figure out what the important factor is. Factor analysis is one technique for discovering the important factor in an effect.

Depending on the predictions, the experiments can have different shapes. It could be a classical experiment in a laboratory setting, a double-blind study or an archaeological excavation. Even taking a plane from New York to Paris is an experiment which tests the aerodynamical hypotheses used for constructing the plane. Scientists assume an attitude of openness and accountability on the part of those conducting an experiment. Detailed record keeping is essential, to aid in recording and reporting on the experimental results, and providing evidence of the effectiveness and integrity of the procedure. They will also assist in reproducing the experimental results. Traces of this tradition can be seen in the work of Hipparchus (190-120 BCE), when determining a value for the precession of the Earth, while controlled experiments can be seen in the works of Muslim scientists such as Geber (721-815 CE), al-Battani (853–929) and Alhacen (965-1039).

DNA-experiments

Watson and Crick showed an initial (and incorrect) proposal for the structure of DNA to a team from Kings College - Rosalind Franklin, Maurice Wilkins, and Raymond Gosling. Franklin immediately spotted the flaws which concerned the water content. Later Watson saw Franklin's detailed X-ray diffraction images which showed an X-shape and confirmed that the structure was helical. This rekindled Watson and Crick's model building and led to the correct structure.

Evaluation and Iteration

Testing and Improvement

The scientific process is iterative. At any stage it is possible that some consideration will lead the scientist to repeat an earlier part of the process. Failure to develop an interesting hypothesis may lead a scientist to re-define the subject they are considering. Failure of a hypothesis to produce interesting and testable predictions may lead to reconsideration of the hypothesis or of the definition of the subject. Failure of the experiment to produce interesting results may lead the scientist to reconsidering the experimental method, the hypothesis or the definition of the subject.

Other scientists may start their own research and enter the process at any stage. They might adopt the characterization and formulate their own hypothesis, or they might adopt the hypothesis and deduce their own predictions. Often the experiment is not done by the person who made the prediction and the characterization is based on experiments done by someone else. Published results of experiments can also serve as a hypothesis predicting their own reproducibility.

DNA-iterations

After considerable fruitless experimentation, being discouraged by their superior from continuing, and numerous false starts, Watson and Crick were able to infer the essential structure of DNA by concrete modeling of the physical shapes of the nucleotides which comprise it. They were guided by the bond lengths which had been deduced by Linus Pauling and by Rosalind Franklin's X-ray diffraction images.

Confirmation

Science is a social enterprise, and scientific work tends to be accepted by the community when it has been confirmed. Crucially, experimental and theoretical results must be reproduced by others within the science community. Researchers have given their lives for this vision; Georg Wilhelm Richmann was killed by ball lightning (1753) when attempting to replicate the 1752 kite-flying experiment of Benjamin Franklin. To protect against bad science and fraudulent data, government research granting agencies like NSF and science journals like *Nature* and *Science* have a policy that researchers must archive their data and methods so other researchers can access it, test the data and methods and build on the research that has gone before. Scientific data archiving can be done at a number of national archives in the U.S. or in the World Data Center.

Models of Scientific Inquiry

Classical Model

The classical model of scientific inquiry derives from Aristotle, who distinguished the forms of approximate and exact reasoning, set out the threefold scheme of abductive, deductive, and inductive inference, and also treated the compound forms such as reasoning by analogy.

Pragmatic Model

Charles Peirce (1839-1914) considered scientific inquiry to be a species of the genus *inquiry*, which he defined as any means of fixing belief, that is, any means of arriving at a settled opinion on a matter in question. He observed that inquiry in general begins with a state of uncertainty and moves toward a state of certainty, sufficient at least to terminate the inquiry for the time being. In 1877, he outlined four methods for the settlement of doubt, graded by their success in achieving a sound fixation of belief:

1. The method of tenacity — persisting in that which one is inclined to think.
2. The method of authority — conformity to a source of ready-made beliefs.

3. The method of congruity or the a priori or the dilettante or "what is agreeable to reason" – leading to argumentation that gets finally nowhere.
4. The scientific method.

Peirce held that slow and stumbling ratiocination can be dangerously inferior to instinct, sentiment, and tradition in practical matters, and that the scientific method is best suited to theoretical research, which in turn should not be trammeled by the other methods and practical ends. What recommends the specifically scientific method of inquiry above all others is the fact that it is deliberately designed to arrive, eventually, at the ultimately most secure beliefs, upon which the most successful actions can eventually be based.

In Peirce's view, the conception of inquiry depends on, but also informs, the conceptions of truth and the real; to reason is to presuppose, as a principle of the reasoner's self-regulation, that the truth is discoverable and independent of our vagaries of opinion. As a question of presuppositions of reasoning, he defined truth as the correspondence of a sign (in particular, a proposition) to its object and, pragmatically, not as any actual consensus of any finite community (i.e., such that to inquire would be to go ask the experts for the answers), but instead as that ideal final opinion which all reasonable scientific intelligences would reach despite individual vagaries, sooner or later but still inevitably, if they pushed investigation far enough.

In tandem he defined the real as a true sign's object (be that object a possibility or quality, or an actuality or brute fact, or a necessity or norm or law), which is what it is independently of any finite community's opinion and, pragmatically, has dependence only on the ideal final opinion. That is an opinion as far or nears as the truth itself to you or me or any finite community of minds. Thus his theory of inquiry boils down to "do the science." At the same time those conceptions of truth and the real involve the idea of a community, both without definite limits and capable of definite increase of knowledge. As inference, "logic is rooted in the social principle".

Paying special attention to the generation of explanations, Peirce outlined scientific method as a collaboration of kinds of inference in a purposeful cycle aimed at settling doubts, as follows:

1. Abduction (or retroduction) - Guessing, inference to the best explanation, generation of explanatory hypothesis. From abduction, Peirce distinguishes induction as inferring, on the basis of tests, the proportion of truth in the hypothesis. Every inquiry, whether into ideas, brute facts, or norms and laws, arises as a result of surprising observations in the given realm or realms, and the pondering of the phenomenon in all its aspects in the attempt to resolve the wonder. All explanatory content of theories is reached by way of abduction, the most insecure among modes of inference. Induction as a process is far too slow for that job, so economy of research demands and even governs abduction, whose modicum of success depends on one's being somehow attuned to nature, by dispositions learned and, some of them, likely inborn.

   Abduction has general justification inductively in that it works often enough and that nothing else works, at least not quickly enough when science is already properly rather slow the work of indefinitely many generations. Peirce calls his pragmatism "the logic of abduction". His Pragmatic Maxim is: "Consider what effects that might conceivably have practical bearings you conceive the objects of your conception to have. Then, your conception of those effects is the whole of your conception of the object". His pragmatism is a method of sorting out conceptual confusions by equating the meaning of any concept with the conceivable practical consequences of whatever it is which the concept portrays.

It is a method of experimentational mental reflection arriving at conceptions in terms of conceivable confirmatory and disconfirmatory circumstances — a method hospitable to the generation of explanatory hypotheses, and conducive to the employment and improvement of verification to test the truth of putative knowledge. Given abduction's dependence on mental processes not necessarily conscious and deliberate but, in any case, attuned to nature, and given abduction's being driven by the need to economize the inquiry process, its explanatory hypotheses should be optimally simple in the sense of "natural" (for which Peirce cites Galileo and which Peirce distinguishes from "logically simple"). Given abduction's insecurity, it should have consequences with conceivable practical bearing leading at least to mental tests, and, in science, lending them to scientific testing.

2. Deduction- Analysis of hypothesis and deduction of its consequences in order to test the hypothesis. Two stages:
    a. Explication-Logical analysis of the hypothesis in order to render it as distinct as possible.
    b. Demonstration (or deductive argumentation)- Deduction of hypothesis's consequence. Corollarial or, if needed, Theorematic.
3. Induction- The long-run validity of the rule of induction is deducible from the principle (presupposition to reasoning in general) that the real is only the object of the final opinion to which adequate investigation would lead In other words, if there were something to which an inductive process involving ongoing tests or observations would *never* lead, then that thing would not be real. Three stages:
    a. Classification-Classing objects of experience under general ideas.
    b. Probation (or direct Inductive Argumentation)- Crude (the enumeration of instances) or Gradual

(new estimate of proportion of truth in the hypothesis after each test). Gradual Induction is Qualitative or Quantitative; if Quantitative, then dependent on measurements, or on statistics, or on countings.

c. Sentential Induction.-"...which, by Inductive reasonings, appraises the different Probations singly, and then their combinations, then makes self-appraisal of these very appraisals themselves, and passes final judgment on the whole result".

Computational Approaches

Many subspecialties of applied logic and computer science, such as artificial intelligence, machine learning, computational learning theory, inferential statistics, and knowledge representation, are concerned with setting out computational, logical, and statistical frameworks for the various types of inference involved in scientific inquiry. In particular, they contribute hypothesis formation, logical deduction, and empirical testing. Some of these applications draw on measures of complexity from algorithmic information theory to guide the making of predictions from prior distributions of experience, for example, see the complexity measure called the *speed prior* from which a computable strategy for optimal inductive reasoning can be derived.

Philosophy and Sociology of Science

Philosophy of science looks at the underpinning logic of the scientific method, at what separates science from non-science, and the ethic that is implicit in science. Philosophy is at least implicitly at the core of every decision we make or position we take, it is obvious that correct philosophy is a necessity for scientific inquiry to take place. There are basic assumptions derived from philosophy that form the base of the scientific method - namely, that reality is objective and consistent, that humans have the capacity to perceive reality accurately, and that rational explanations exist for elements of the real world. These assumptions from metaphysical naturalism form the basis on which science is grounded.

We find ourselves in a world that is not directly understandable. We find that we sometimes disagree with others as to the facts of the things we see in the world around us, and we find that there are things in the world that are at odds with our present understanding. The scientific method attempts to provide a way in which we can reach agreement and understanding. A "perfect" scientific method might work in such a way that rational application of the method would always result in agreement and understanding; a perfect method would arguably be algorithmic, and so not leave any room for rational agents to disagree.

As with all philosophical topics, the search has been neither straightforward nor simple. Logical Positivist, empiricist, falsificationist, and other theories have claimed to give a definitive account of the logic of science, but each has in turn been criticized. Thomas Samuel Kuhn examined the history of science in his *The Structure of Scientific Revolutions*, and found that the actual method used by scientists differed dramatically from the then-espoused method. His observations of science practice are essentially sociological and do not speak to how science is or can be practiced in other times and other cultures. Imre Lakatos and Thomas Kuhn have done extensive work on the "theory laden" character of observation. Kuhn (1961) said the scientist generally has a theory in mind before designing and undertaking experiments so as to make empirical observations, and that the "route from theory to measurement can almost never be traveled backward".

This implies that the way in which theory is tested is dictated by the nature of the theory itself, which led Kuhn to argue that "once it has been adopted by a profession ... no theory is recognized to be testable by any quantitative tests that it has not already passed". Paul Feyerabend similarly examined the history of science, and was led to deny that science is genuinely a methodological process. In his book *Against Method* he argues that scientific progress is *not* the result of applying any particular method. In essence, he says that "anything goes", by which he meant that for any specific methodology or norm of science, successful science has been done in violation of it. Criticisms such as led to the strong programme, a radical approach to the sociology of science.

In his 1958 book, *Personal Knowledge,* chemist and philosopher Michael Polanyi (1891-1976) criticized the common view that the scientific method is purely objective and generates objective knowledge. Polanyi cast this view as a misunderstanding of the scientific method and of the nature of scientific inquiry, generally. He argued that scientists do and must follow personal passions in appraising facts and in determining which scientific questions to investigate. He concluded that a structure of liberty is essential for the advancement of science - that the freedom to pursue science for its own sake is a prerequisite for the production of knowledge through peer review and the scientific method.

The post-modernist critiques of science have themselves been the subject of intense controversy. This ongoing debate, known as the science wars, is the result of conflicting values and assumptions between the postmodernist and realist camps. Whereas postmodernists assert that scientific knowledge is simply another discourse (note that this term has special meaning in this context) and not representative of any form of fundamental truth, realists in the scientific community maintain that scientific knowledge does reveal real and fundamental truths about reality. Many books have been written by scientists who take on this problem and challenge the assertions of the postmodernists while defending science as a legitimate method of deriving truth.

Communication, Community, Culture

Frequently the scientific method is not employed by a single person, but by several people cooperating directly or indirectly. Such cooperation can be regarded as one of the defining elements of a scientific community. Various techniques have been developed to ensure the integrity of the scientific method within such an environment.

Peer Review Evaluation

Scientific journals use a process of *peer review*, in which scientists' manuscripts are submitted by editors of scientific journals to (usually one to three) fellow (usually anonymous) scientists familiar with the field for evaluation. The referees may or may not recommend publication, publication with suggested modifications, or, sometimes, publication in another journal. This serves to keep the scientific literature free of unscientific or crackpot work helps to cut down on obvious errors, and generally otherwise improve the quality of the scientific literature.

Documentation and Replication

Sometimes experimenters may make systematic errors during their experiments, unconsciously veer from the scientific method (Pathological science) for various reasons, or, in rare cases, deliberately falsify their results. Consequently, it is a common practice for other scientists to attempt to repeat the experiments in order to duplicate the results, thus further validating the hypothesis.

Archiving

As a result, researchers are expected to practice scientific data archiving in compliance with the policies of government funding agencies and scientific journals. Detailed records of their experimental procedures, raw data, statistical analyses and source code are preserved in order to provide evidence of the effectiveness and integrity of the procedure and assist in reproduction. These procedural records may also assist in the conception of new experiments to test the hypothesis, and may prove useful to engineers who might examine the potential practical applications of a discovery.

Data Sharing

When additional information is needed before a study can be reproduced, the author of the study is expected to provide it promptly - although a small charge may apply. If the author refuses to share data, appeals can be made to the journal editors who published the study or to the institution which funded the research.

Limitations

Note that it is not possible for a scientist to record *everything* that took place in an experiment. He must select the facts he believes to be relevant to the experiment and report them. This may lead, unavoidably, to problems later if some supposedly irrelevant feature is questioned. For example, Heinrich Hertz did not report the size of the room used to test Maxwell's equations, which later turned out to account for a small deviation in the results. The problem is that parts of the theory itself need to be assumed in order to select and report the experimental conditions. The observations are hence sometimes described as being 'theory-laden'.

Dimensions of Practice

The primary constraints on contemporary western science are:

- Publication, i.e. Peer review
- Resources (mostly funding)

It has not always been like this: in the old days of the "gentleman scientist" funding (and to a lesser extent publication) were far weaker constraints. Both of these constraints indirectly bring in a scientific method — work that too obviously violates the constraints will be difficult to publish and difficult to get funded. Journals do not require submitted papers to conform to anything more specific than "good scientific practice" and this is mostly enforced by peer review. Originality, importance and interest are more important - see for example the author guidelines for *Nature.*

Criticisms of these restraints are that they are so nebulous in definition (e.g. "good scientific practice") and open to ideological, or even political, manipulation apart from a rigorous practice of a scientific method that they often serve to censor rather than promote scientific discovery. Apparent censorship through refusal to publish ideas unpopular with mainstream scientists (unpopular because of ideological reasons and/or because they seem to contradict long held scientific theories) has soured the popular perception of scientists as being neutral or seekers of truth and often denigrated popular perception of science as a whole.

The development of the scientific method is inseparable from the history of science itself. Ancient Egyptian documents, such as early papyri, describe methods of medical diagnosis. In ancient Greek culture, the method of empiricism was described. The first experimental scientific method was developed by Muslim scientists, who introduced the use of experimentation and quantification to distinguish between competing scientific theories set within a generally empirical orientation, which emerged with Alhazen's optical experiments in his *Book of Optics* (1021). The modern scientific method crystallized no later than in the 17th and 18th centuries. In his work *Novum Organum* (1620) — a reference to Aristotle's *Organon* — Francis Bacon outlined a new system of logic to improve upon the old philosophical process of syllogism.

Then, in 1637, René Descartes established the framework for a scientific method's guiding principles in his treatise, *Discourse on Method.* The writings of Alhazen, Bacon and Descartes are considered critical in the historical development of the modern scientific method. In the late 19th century, Charles Sanders Peirce proposed a schema that would turn out to have considerable influence in the development of current scientific method generally. Peirce accelerated the progress on several fronts. Firstly, speaking in broader context in "How to Make Our Ideas Clear" (1878), Peirce outlined an objectively verifiable method to test the truth of putative knowledge on a way that goes beyond mere foundational alternatives, focusing upon both *deduction* and *induction*.

He thus placed induction and deduction in a complementary rather than competitive context (the latter of which had been the primary trend at least since David Hume, who wrote in the mid-to-late 18th century). Secondly, and of more direct importance to modern method, Peirce put forth the basic schema for hypothesis/testing that continues to prevail today. Extracting the theory of inquiry from its raw materials in classical logic, he refined it in parallel with the early development of symbolic logic to address the then-current problems in scientific reasoning. Peirce examined and articulated the three fundamental modes of reasoning that, as discussed above in this chapter, play a role in inquiry today, the processes that are currently known as abductive, deductive, and inductive inference. Thirdly, he played a major role in the progress of symbolic logic itself — indeed this was his primary specialty.

Karl Popper denied the existence of evidence and of scientific method. Popper holds that there is only one universal method, the negative method of trial and error. It covers not only all products of the human mind, including science, mathematics, philosophy, art and so on, but also the evolution of life. Beginning in the 1930s and with increased vigor after World War II, he argued that a hypothesis must be falsifiable and, following Peirce and others, that science would best progress using deductive reasoning as its primary emphasis, known as critical rationalism. His formulations of logical procedure helped to rein in excessive use of inductive speculation upon inductive speculation, and also strengthened the conceptual foundation for today's peer review procedures.

Relationship with Mathematics

Science is the process of gathering, comparing, and evaluating proposed models against observables. A model can be a simulation, mathematical or chemical formula, or set of proposed steps. Science is like mathematics in that researchers in both disciplines can clearly distinguish what is *known* from what is *unknown* at each stage of discovery. Models, in both science and mathematics, need to be internally consistent and also ought to be *falsifiable* (capable of disproof). In mathematics, a statement need not yet be proven; at such a stage, that statement would be called a conjecture. But when a statement has attained mathematical proof, that statement gains a kind of immortality which is highly prized by mathematicians, and for which some mathematicians devote their lives.

Mathematical work and scientific work can inspire each other. For example, the technical concept of time arose in science, and timelessness was a hallmark of a mathematical topic. But today, the Poincaré conjecture has been proven using time as a mathematical concept in which objects can flow. Nevertheless, the connection between mathematics and reality (and so science to the extent it describes reality) remains obscure. Eugene Wigner's paper, *The Unreasonable Effectiveness of Mathematics in the Natural Sciences*, is a very well-known account of the issue from a Nobel Prize physicist.

In fact, some observers (including some well known mathematicians such as Gregory Chaitin, and others such as Lakoff and Nunez) have suggested that mathematics is the result of practitioner bias and human limitation (including cultural ones), somewhat like the post-modernist view of science. George Pólya's work on problem solving, of the construction of mathematical proofs and heuristic show that mathematical method and scientific method differ in detail, while nevertheless resembling each other in using iterative or recursive steps.

Mathematical method	*Scientific method*
1. Understanding	Characterization from experience and observation
2. Analysis	Hypothesis: a proposed explanation
3. Synthesis	Deduction: prediction from the hypothesis
4. Review/Extend	Test and experiment

In Pólya's view, *understanding* involves restating unfamiliar definitions in your own words, resorting to geometrical figures, and questioning what we know and do not know already; *analysis*, which Pólya takes from Pappus, involves free and heuristic construction of plausible arguments, working backward from the goal, and devising a plan for constructing the proof; *synthesis* is the strict Euclidean exposition of step-by-step details of the proof; *review* involves reconsidering and re-examining the result and the path taken to it.

Scientists and Science

Science is so exciting! Why? Because it is awe-inspiring, fun and creative. Most scientists would not do anything else — they are truly dedicated to what they do. You can hardly get them to be quiet once they start talking about their work. Unlike most people, they usually love their jobs! However, this is not the common view of most scientists. They are commonly thought of as nerds, freaks, weirdoes, or evil doers, but this is a movie or television view. The movies and TV are fantasies where, with a few exceptions, scientists are shown negatively because it fits dramatic needs. It's just not true in real life.

Another misunderstanding about science and scientists comes from the field itself — the words, mathematics, and reasoning may appear convoluted and difficult. Commonly, science is extremely complex and thus unintelligible to the lay person or even other scientists in different fields. Therefore the people doing it must be geniuses. True, our most famous scientist, Einstein, probably was a genius, but just because scientists deal with complex subjects does not mean that they are necessarily any smarter than anyone else. They are trained to think in certain ways, but these ways are attainable by most people. Likewise, science, itself, need not be complicated, although it may be, just like so many other endeavors in our lives. Much science can be easily understood and pursued by lay people and amateurs.

Science and Non-scientists

All this complexity and incomprehensible information must seem very discouraging or esoteric to a non-scientist, yet science is enormously important to every person in our society. All of our technology and many political decisions are concerned with fundamental scientific issues. Everyone needs to understand the process and nature of science, if not some of the facts, simply to live intelligent, productive lives in our modern society. Science impacts our daily lives, our long-term health and well-being, our children, the policies and decisions made that affect our cities, counties, states, and country, and the nature of the world we live in. Understanding the process of science is actually not so difficult.

The basic way of scientific thinking is already practiced informally by almost everyone. It even appears in our language with such phrases as "common sense" and "tried and true." These are ubiquitous and indicate a way of thinking that is commonly scientific. Such phrases suggest that repeated observation and experimentation work for the ordinary person. Everyone does science! You just might not know it. How do we do that? At one level or another, we all think critically about things around us, we usually gather some evidence, and we wonder about the people who tell (or sell) us things. Honed true, this is the basic stuff of science. All scientists practice and use these skills in their daily work. We should all use them in our daily lives too, for they provide the means to a better, more is rewarding, and more fulfilling life— better than just winging it. Practiced all the time, you might even save a good deal of money!

Think back to when you were a kid. You may well have loved animals, rocks, fossils, stars, bugs, clouds, or a multitude of other natural objects. You might have even collected some of them and tried to put them in some kind of order. Maybe you tried to figure out how to throw a baseball better or ski faster or build models that really worked. You might have tried cooking or dress designing or mixing chemicals to see what happened. Maybe you made clay dinosaurs and pushed them around the floor. You were excited; you were having fun; and you were probably using science, but you didn't even know it. Even as an adult, you still look both ways when crossing the street; you still try to figure out how to solve some ordinary problem; and you probably examine you foodstuffs before you buy them. All of these involve critical thinking, evidential reasoning, and evaluation of authority — each a critical element in doing good science.

The objective of doing science is simple — to learn about the things and processes around us, from the expanse of the Universe to our own bodies. Scientists seek explanations that consistently work and that are consistent with all other scientific knowledge. In this sense, it works towards truth, but truth may be more ephemeral. Science uses critical thinking, appropriate evidence, subjects all authority to scrutiny, and allows testing of its claims in order to do this. This may sound dull, but in fact these are easily learned and are powerful skills for everyone to have. For example, if someone tries to sell you something, say a health potion, your skills in these areas may determine whether or not you waste your time and money, endanger your health, or live a longer life. They can help you to evaluate almost any situation or claim that you encounter. You will be free to make informed decisions about your own life rather than depend on others.

Critical Thinking, Evidential Reasoning and Judging Authority

Critical Thinking

This involves eight skills. These skills require that you understand the problem clearly, consider all possible views about the problem, set emotion aside, and be willing to be flexible when solutions are imperfect. The skills will aid you in dealing with the problem.

The first three critical skills may be self-evident, but the others are commonly difficult for people to practice because of human nature. The analysis of assumptions and biases requires a certain amount of personal insight. We all have biases based on our past experiences and personal beliefs, but they must be set aside when we need to understand the way the world works. This is often very difficult to do, because we are not even aware of many of our personal biases. One way to identify bias is to make a list of your feelings and knowledge about the subject. Then apply the evidence. If it does not support your feeling, perhaps the feeling is unjustifiable. Later, after examining other factors, you can return to this issue with a better understanding of your own emotional biases. If in conflict, your feelings should probably be suppressed in favor of the evidence.

Table: Skills involved in critical thinking. Simple techniques for achieving these skills are also suggested that should be applied every time you are faced with a problem. Scientists train themselves to use these skills all the time, but even they may occasionally fail to do so.

Skills	*Simple Techniques*
1. Ask questions: be willing to wonder.	Start by asking "Why?"
2. Define the problem.	Restate the issue several different ways so it is clear.
3. Examine the evidence refutes the claim.	Ask what supports or Is it reliable?
4. Analyze assumptions and biases.	List the evidence on which each part of the argument is based. The assumptions and biases will be unsupported and should be eliminated from further consideration.
5. Avoid emotional reasoning.	Identify emotional influence and "gut feelings" in the arguments and exclude them.
6. Don't oversimplify.	Do not generalize from too little evidence.
7. Consider other	Make sure alternate views are included in the discussion.
8. Tolerate uncertainty.	Be ready to accept tentative answers when evidence is incomplete and new answers when further evidence warrants them.

The last three items are particularly difficult. We all want explanations and we tend to jump to conclusions based on too little evidence. Again, an analysis of the evidence is required to determine if it is sufficient. Alternative interpretations should always be sought, even if the evidence seems compelling. Does the evidence allow for other possible interpretations? And lastly, tolerate uncertainty. No one likes uncertainty in their lives — we all want, perhaps need, to know what is before us, why things happen to us, what happens when we die, and many, many more. Although difficult to do, tolerating uncertainty can be done by simply setting aside the uncertainties and, for the moment at least, accepting the uncertain.

Evidential Reasoning

In science, as in our daily lives, various claims are made by other scientists or acquaintances, TV personalities, doctors, and so on. We might even make such claims ourselves. Of these points, perhaps the most critical is the last one. Any claim must be sufficient. In other words, you do not have to prove that the claim is false in order to test it, the claimant must provide sufficient proof himself. Secondly, the more extraordinary a claim, the more extraordinary the evidence must be to test it. For example, if a person claims that some herb has cured their cancer, you would be well advised to seek a good deal of further supporting evidence before risking your own life. Or if a person claims to have an extraterrestrial being in their garage, do not accept a fuzzy or even clear photograph as proof, demand a piece of the thing for further study. And lastly, the word of someone is never sufficient to establish the truth of a claim.

Table: Rules for evidential reasoning, or a guide to intelligent living and the scientific method. All claims, whether scientific or not, should be subjected to these rules in order to ensure that all possibilities are considered fairly.

Rules for Evidential Reasoning	*What to do*
1. Falsifiability	Conceive of all evidence that would prove the claim false.
2. Logic	Argument must be sound.
3. Comprehensiveness	Must use all the available evidence.
4. Honesty	Evaluate evidence without self-deception.
5. Replicability	Evidence must be repeatable.
6. Sufficiency	A. Burden of proof rests on the claimant. B. Extraordinary claims require extraordinary evidence. C. Authority and/or testimony is always inadequate.

Judging Authority

The evaluation of authority requires special consideration. We often need to rely on other people for certain kinds of information or decisions we make in science or in our lives, simply because we cannot know enough about everything. Who can we trust to provide us with solid information?

Even when an authority passes these tests, be aware of lapses that may reveal the degree of knowledge possessed by an expert. Well known or highly honored scientists are commonly asked to comment on subjects outside their own field of expertise. These people should be subjected to exactly the same questions as an unknown authority. Does a Noble prize winner in physics, for example, have any credibility when making pronouncements about evolution? Maybe, but the evidence and hypotheses about evolution are very far removed from the usual literature of physics. Be suspicious. In your daily life, authorities are always present making one claim or another, usually to sell something. These claimants can be evaluated rationally too. So critical thinking, evidential reasoning and judging authority are essential to the scientific process, as well as to living an intelligent, full, and good life. Science requires more, however, and that is the development of knowledge.

Scientific Knowledge

Science is a Social Activity

Table: How to evaluate authority.

Criterion	What to Consider	What to Do
1) Does the authority have proper credentials?	Considerable study or experience in a subject is required to become an expert in any field. Does the authority have degrees from a recognized college or university that has the faculty, libraries, and other facilities for proper education in the subject? Has the authority worked in the field for some time?	If not, don't believe him.
2) Does the authority have proper affiliations?	Is she identified closely with a reliable organization, such as a university, museum, government agency or corporation that practices the subject?	If not, ask how she makes a living?
3) Does that organization have a stake in the claims made by the claimant?	Be suspicious of anyone making claims that support the position or products of their own organization.	Seek independent evidence that the claim is correct.
4) Has the authority subjected her work to peer review?	In other words, have other experts evaluated the work, so that some independent assessment has been made?	If not, seek that evaluation yourself or find another authority.
5) Does the authority use the skills of critical thinking and evidential reasoning?	Does he use items listed in Tables 1 and 2?	If not, question him using those very skills yourself.
6) Is the authority a demonstrated expert in their field?	Other trustworthy people should rely on this person's expertise. Do other experts cite her conclusions?	If not, find another authority who others do rely on.
7) Does the authority present arguments without undue call on unsupported or untenable claims?	Does the authority present evidence sufficient that you can evaluate it?	If not, find an authority who can provide evidence supporting the claims.

Scientific knowledge results from the efforts of hundreds of thousands of people over hundreds of years. Even current scientific efforts on single problems usually involve many people working in different ways all over the world. Scientific investigation is a group effort, requiring communication among all workers. In this way, all interested scientists can examine the work and conclusions of other scientists. Science is done in a social context.

The Scientific Knowledge Base is Huge. Few scientists (probably none) can speak intelligently about all aspects of science because it is so enormous. They are usually experts in only one small field of a larger discipline, just like so many other fields of human endeavor. Some may speak with authority on other aspects of science but they should be regarded with caution and further analysis. Because scientists differ in no significant respect from other people, they have the same strengths and weaknesses, and this may show in their work. All scientists are especially proud of their own ideas, for example, and they are ready to defend them. A good scientist will change her views, if enough evidence is accumulated to counter them. Some never change their minds, even when incontrovertible evidence appears. That is simply human nature, not a failure of science.

Controversy in Science

Controversy abounds in science and can be extremely helpful — controversy is not a weakness. Controversy, in fact, is almost certain to develop in science because the testing of hypotheses is devised to disprove not prove them. Thus, everyone tries to disprove another's hypotheses and that scientist then may refute that disproof with yet additional evidence or presentation of alternative hypotheses. In this process of controversy, science advances. Because science is actually a group activity, each scientist builds on the work of others through criticism and support, and then contributes ideas to be used or criticized by those who follow. This is one of the great satisfactions that scientists feel — their own ideas may last an eternity! But even a bad idea is often more constructive in the long run, because other scientists may sense that it is wrong and seek vigorously to correct it. A lingering bad idea does not indicate that science is lacking, but simply that someone has not yet taken it to task.

Empirical and Historical Science

The sciences are not all the same either. Although the process of science remains the same, the nature of the observations may differ. These observations can be either non-historical (time independent) or historical (time-dependent). Sciences like physics, chemistry and much of molecular biology are largely non-historical, although they may rely on historical observations in particular instances. They deal with observations that are not expected to change with time — they are time independent. An experiment done today should produce the same results as one done 10 years ago or 10 years in the future. For example, water should always flow downhill because the effect of gravity does not change with time, whether it is a billion years ago, yesterday, or today.

We can expect that gravity will not change in another billion years. Sciences like astronomy, anthropology, much of biology, geology, paleontology, and evolutionary biology are largely historical, although each uses experiments commonly. Historical sciences rely on observations or evidence (results) of phenomena that happened in the past. These results arose through a series of events, the history, and each event was contingent on previous ones. Historical scientists can only infer the causes from the results, since the results happened in the past. Many ordinary objects and processes around you have strong historical backgrounds, including your own life. You are a result of an historical process which cannot be directly observed.

A good geologic example is the scientific study of the Grand Canyon. We see the Canyon today because the region has experienced a series of events in a particular order. If the events or order were different, the Grand Canyon might not exist as it is. If the metamorphism of the Vishnu Schist now at the bottom of the Canyon or the deposition of sand, mud and lime in the Paleozoic seas that once covered the area had not occurred, then the Grand Canyon would look quite different, if it would exist at all. The historical sciences, of course, also use experimental methods. The rate of erosion of the rocks in the Grand Canyon can be determined by repeatable experiments, and thus provide additional evidence about how the Canyon came to be.

By accumulating multiple observations throughout the Grand Canyon and the Colorado Plateau in which it is cut and conducting critical experiments, geologists have reconstructed a reliable model for the formation of the Grand Canyon. A similar process takes place in the study of the evolution of humans, rats, rice, or any other organism. Historical science has a greater margin of error most of the time than non-historical, experimental science because scientists cannot repeat each event and must view only the results of those events through a filter of time. That margin should not, however, be mistaken for a lack of knowledge. We understand the formation of the Grand Canyon in all aspects, but not in every detail.

In evolutionary biology, so-called missing links are details, not evidence that destroys the theory. Just because you may not have any information about your great great grandmother does not invalidate her existence nor is that she a part of your history. Paleontology and evolutionary biology are largely historical sciences that reveal the broad patterns, and very commonly even the detailed patterns, of evolutionary history. Gaps do exist in that record, just as there are likely to be gaps in your own family's historical record, but that does not invalidate or make the science less substantive.

Scientists and teachers tout the "scientific method" as the way we do science; it appears dull and agonizing. This erroneous formalization of the scientific thinking process makes people fear science. Few scientists actually work that way. Instead, they get excited, hopeful, interested, intrigued, and puzzled. Their ideas come to them in the shower, on the freeway, while playing baseball with their kids, as well as in the laboratory, office or library. Good science is a creative process without rules on how to do it.

Alternative Ways of Doing Science

How, then, do scientists sort out all this creativity to make order and understanding possible? They use various ways. More than a hundred years ago, T.C. Chamberlain laid out three ways of doing science. Chamberlain and later authors made clear that the third method, that of multiple working hypotheses, was the most efficient and preferred way to proceed. Scientists do three fundamental activities. They gather data, develop hypotheses, and test hypotheses, not necessarily in that order. Within these activities, individual scientists may work differently. Chamberlain (1897) and Platt (1964) carefully pointed out the strengths and weaknesses of each method. Data gathering is simply the accumulation of observations — experimental, observational, or mathematical. Hypothesis development requires that it be consistent with all known data, that it fit logically within other accepted hypotheses in science, that it be testable, and that it have predictive power.

Hypotheses must be testable within the abilities of science; otherwise they remain simply an idea without use. They must also predict that certain phenomena will occur if certain experiments or observations are undertaken. A test must attempt to disprove the hypothesis since proof in science cannot be attained. The more critical tests that an hypothesis passes, the more confidence we can have in it. The concept of a scientific hypothesis differs significantly from the common use of "theory" in non-scientific language.

"Theory" here means an ad hoc idea which does not possess the characteristics of a good hypothesis. The common usage of "theory" should not be confused with the scientific use of "hypothesis" or "theory."

Table: Ways of doing science, according to Chamberlain (1897).

Type of Approach	What is Done	Good Points	Bad Points
1. Description of phenomena	Describes what is observed.	Provides necessary data.	Fails to provide explanation.
2. Single Hypothesis method to account for phenomenon.	Provides a single, testable hypothesis.	Scientist may devote much energy and creativity to "proving" or supporting her/his idea	Other hypotheses may equally or better explain the phenomenon . Scientist may be blinded to these other hypotheses.
3. Method of Multiple Working Hypotheses all worked nearly simultaneously .	Provides several too many possible hypotheses for testing with the available data.	Many ideas are tested with the same or little additional effort, and new hypotheses can be generated as they develop. Scientist is not wed to a favorite idea, hence is less biased.	Energy may be dissipated among many hypotheses.

Descriptive Science

Descriptive science is important and nearly all scientists engage in it. It provides the fundamental data surrounding any scientific problem. However, simply answering the "what" questions, without understanding the "why" questions, is unsatisfactory, although necessary. Description of observations is essential to the understanding of processes, for a scientist must know what he is trying to explain. But mere description by itself does little to explain what the processes are that have taken place to give rise to the observations. Observations in science include an experiment, say in chemistry, and of objects in context, say fossils in a rock layer.

An experiment is simply a single observation on a par with an observation of those fossils in rock. Both observations take on significance when they can be made multiple times, thus confirming their generality. The results of an experiment can be readily measured with precision and commonly repeated, but it is nevertheless just another fact, as are the fossils (for which precise measurements and repeated observations are common), supporting or disproving a scientific hypothesis. Although many scientists must stop, at least at times, with description, most proceed to one of the next steps.

The Single Hypothesis Method

A single idea, model or hypothesis to explain a set of observations is commonly developed by scientists. A large number of scientists stop with the development of a hypothesis that accounts for their observations. However, the method of using single hypotheses is fraught with many pitfalls. First, a scientist with a single hypothesis is a knight in armor. He is forced to defend his idea because it is the only one that he possesses. Second, data that do not fit the hypothesis are easy to ignore because there is no other place to use it. Thus, the data collected tend to support the hypothesis, yet the best supported hypothesis can still fail on a single critical observation. Third, a scientist with a single hypothesis has her ego at stake, and thus resists counter hypotheses made by other scientists. Because scientists are like other people with regard to their egos, this resistance to alternate hypotheses results in a loss of objectivity, and sometimes bitterness may ensue and controversy abound when others try to disprove the hypothesis.

The Method of Multiple Working Hypotheses

A better way for scientists to work, and many do, is to use Chamberlain's "Method of Multiple Working Hypotheses". In this method, a scientist thinks of all the possible hypotheses that might account for his or her observations, and then goes on to test each one. In this way, his ego is attached not to a single hypothesis, but to the development and testing of all of them. She takes pride not in a single hypothesis but in an array of them. She has an easier time using all the data, for many hypotheses are available to apply them to. Furthermore, single tests can be devised for each hypothesis and those that fail can be eliminated rapidly, thus increasing the efficiency of scientific progress. For those hypotheses that survive all conceivable tests, they can be arranged in some order of more probable to less probable based on the degree of support generated for each from the data, and the predictive value of the hypothesis.

This arrangement then provides a research plan for other or future workers who may have additional data or newer techniques or instruments. In applied situations where decisions must be made on incomplete information, for example your health or environmental issues, a scientist left with two or three hypotheses can still make an informed decision based on which one is best supported by the available evidence. Science is thus more efficient when the method of "multiple working hypotheses" is fully utilized. To show in a simple way, how each might be applied to a hypothetical study, assume that three investigators want to know why a certain butterfly species congregates in certain kinds of trees. A descriptive scientist might spend the rest of his life counting butterflies in numerous trees, and know only that each tree has some capacity for butterflies without understanding why. While that might be important, it is not an efficient methodology.

A second scientist might have a single favorite hypothesis, say that the butterflies need certain nectar from the trees' flowers and that is why the butterflies congregate there. This scientist will work hard and long to find ways to prove his assertion, for he has no alternatives except to try to prove his idea. A third scientist, using the method of multiple working hypotheses, will acquire some data, develop a number of hypotheses to account for those observations, then design tests that will disprove (not prove) each hypothesis. She may find a single tree someplace that contains an anomalous number of butterflies, thus suggesting that the first scientist has wasted a good deal of his life trying to understand numbers in trees. She may find that some trees with butterflies have no nectar, thus disproving the favorite hypothesis of the second scientist with a single, well understood observation.

A single failed test means that hypothesis can be discarded. In this process, additional data are collected that can be brought to bear on the remaining hypotheses, as well as the development of yet additional ones. A large number of possibilities are dealt with efficiently and quickly. The third scientist moves through all previously proposed hypotheses and those she developed herself, eliminating some and accumulating evidence to support those remaining. In the end, since no hypothesis can ever be "proved", she has several that are supported in one degree or another. The scientist can then develop further tests, or communicate the work to others for them to consider. Science has thus progressed quickly, efficiently, and a future research plan has been laid out, and with these has come greater understanding of the phenomenon, although not necessarily the "truth". We may never know when truth is attained, for to prove scientific hypotheses is impossible.

Science is not based on Faith, but on Hypotheses Testing

Occasionally someone asserts that science is based on faith. This is simply not true. Someone knowledgeable in the scientific process accepts the proclamations of other scientists outside their own field of expertise because we know that the scientific process yields testable hypotheses subject to elimination. The proclamation may not be correct but we accept it anyway, knowing that it is subject to scrutiny and correction by other scientists more expert in that particular field. As a paleontologist, for instance, I tentatively accept Einstein's Theory of Relativity, not on the basis of faith, as some people would assert, but on the knowledge that it has been tested, will continue to be tested, and that it has withstood those tests. Faith is not involved at all. Some hypotheses and theories may not yet have been subjected to the critical tests, either because no other scientist has gotten interested in the problem or the technology or proper discoveries have not yet developed. In this sense, science is self-correcting and errors are eliminated. They may not be eliminated immediately, but eventually they will be. In rapidly moving fields, like molecular biology, corrections come every day; in other less active fields, say sponge systematic, corrections may take decades.

Evolution is the scientific study of how organisms — bacteria, archaea, protists, fungi, animals and plants — came to be. Evolution relies on the same sort of scientific processes and evidence as do all other fields of science. The methods and processes of evolutionary biology do not differ in any significant way from those used in any other science. Evolutionary biologists use critical thinking, evidential reasoning, and judgment of authority, hypotheses development, data gathering, and hypotheses testing just like all other scientists. Evolutionary biologists practice non historical or historical science or both. All biologists contribute to evolution in one way or another. Some use experiments, for example with fruit flies, while others observe the sequence of fossilized organisms in the rock record. Some rely heavily on the other sciences for complementary data, for example the evolutionary paleontologist is beholden to physicists for the understanding of dating by radioactive decay of various rocks.

Darwin brought together a large number of observations and developed the hypothesis that the environment acted to select individual variants in a population that had particular characteristics. He did not know what caused the variation, such as eye color, growth rates, or height, but he determined that if a particular character were consistently selected by the environment through successful breeding, in the same way that dog breeders selected characters they desired in their dogs, then changes would ensue in future generations. This hypothesis he called "natural selection".

As he and others gathered more evidence and added additional hypotheses in the following decades, his hypothesis became strengthened and called a theory. Evolution is a simple elegant theory in its basics, yet it rapidly becomes complex. This complexity leads to other hypotheses to account for aspects of the overall theory. For example, punctuated equilibrium — the idea that species are in some kind of evolutionary stasis for long periods of geologic time and then rapidly evolve into another species — was presented as an alternative to Darwin's idea that evolution was a gradual affair with changes taking place slowly and evenly over a long time. Punctuated equilibrium enhanced Darwin's theory by clarifying how evolution took place through time — it did not disprove the theory as some religious zealots have claimed. Today, many biologists and paleontologists around the world are working to better understand and test Darwin's theory in all its details. Some of these biologists confirm Darwin's ideas, some develop more critical details, and all constantly test the theory, but the theory of evolution has yet to be disproved in spite of all this effort.

Science is a way to understand our world and universe through repeated reliable observations and hypothesis development and testing. The observations must be repeatable by others besides the originator. From this, we conclude that nature is independent of a single observer or set of observers. Nature behaves similarly under similar conditions. It does not differ from observer to observer, as long as the conditions are identical. Thus, science is reliable; observations and tests are not dependent on the kind of people who practice it, what they might feel about it, or what their previous experiences are. Anyone using the same techniques and procedures will have similar or identical results. As far as we can tell, the world and universe are real.

Science requires certain skills — the ability to think critically, to reason evidentially, and to evaluate authority. Using these skills, scientists can then gather data, develop hypotheses to account for the data, and test the hypotheses. The most efficient method of advancing science is the "Method of Multiple Working Hypotheses" that entails simultaneous and continuous development and testing of a number of hypotheses. Hypotheses and theories in science are never proven, only disproven. In this sense, a final, a true answer may never be attained, but the supporting evidence, resistance to disproof, and the logical fit with other scientific knowledge provides differing degrees of confidence. By evaluating the degree of confidence, scientists can make valuable decisions about scientific issues and design new technologies even if an answer is not known with certainty.

In order to do this well, scientists must be excited, creative, and innovative. The best scientists are the most creative scientists. They let their minds wander, putting various ideas and bits of data together to find possible solutions to a particular puzzle about the universe. Look at the pictures from the Hubble Space Telescope, the images from the Martian Pathfinder, or the bones of a dinosaur partially excavated from the ground. Imagine the thrill of waiting for an experiment to run to completion, of studying a whale breaching at the ocean's surface, a comet hitting a planet, or a snake eating a rat. These are each different and attract different kinds of people, but they can all be good science. This is excitement, and it is addictive! In a good way and often for life. A good life. Science is fun! And the good news is: Almost anyone can learn and practice the skills of science and improve their lives.

7

From Applied Science to Techno-Science

As soon as the new sciences of Complexity, Chaos Theory, and Artificial Life (hereafter referred to as AL), began to be noticed by the popular science press a kind of "honeymoon" period began. During this time these sciences were seen as the sexy new breakthrough theories that would eventually lead to our ability to solve all the problems of the world, from the cure for AIDS to the complete understanding and synthesis of living systems. Recently a number of attacks have been leveled against the studies of Complexity and Chaos Theory in general and on the study of AL directly. The most damning of these attacks on AL has been launched by John Horgan in his article "From Complexity to Perplexity," printed in Scientific American and in his book *The End of Science*. In his article Horgan fiercely criticized the study of AL with the implication that the entire study is some kind of sham. Horgan states that:

Artificial Life — and the entire field of complexity — seem to be based on a seductive syllogism: There are simple sets of mathematical rules that when followed by a computer give rise to extremely complicated patterns. The world also contains many extremely complicated patterns. Conclusion: Simple rules underlie many extremely complicated phenomena in the world. With the help of powerful computers, scientists can root those rules out.

Horgan goes on to argue that this position held by AL researchers is untenable and that the kind of science practiced in AL is no where near the kind of science that is normally practiced in biology. Mathematical models are not as accurate as one would like when studying natural systems, in fact presented arguments myself which attempt to refute some of the strong claims made by AL researchers as to the ability of mathematics to embody living systems. But I think that Horgan takes his argument too far. Horgan suggests that the synthetic nature of the study of AL is closer to poetry than it is to science.

Horgan seems to feel that this fact degrades AL in comparison to other sciences. One can only hold a position such as Horgan's if one fails to understand that AL is an example of a techno science as are all modern sciences, and that synthesis is and always has been a part of these knowledge producing endeavors. In order to prove this point we will go through some recent theories presented in the literature of the field of the philosophy of technology where we will look at the implications of how science relates to technology and how both effect the societies in which they are contained. Once this is accomplished we will see that it is not AL that is fundamentally flawed, rather, our problems rest in our old understanding of the relationship between science and technology.

Artificial Life: A Technology or a Science

When one thinks about the subject of Artificial Life it is hard to determine if AL is the study of a new science of life like systems or if it is the practice of a specialized subset of computer technology. The problem arises because it is not possible to imagine the current study of AL without the use of advanced computer technology. Artificial Life is the attempt to model and synthesize life in a dynamic and complex way (Langton 1989). As it turns out the study of complex dynamic systems which exhibit a complexity resembling that of a living system is best accomplished by conceiving of a mathematical model and then letting it run on a suitably programmed computer. The modeled system's behavior can then be analyzed, allowing the researcher to then make a judgment about the model's similarity to a truly living system.

These complex models can be very taxing on computer resources and serious AL research can only be accomplished on powerful computing machinery. For this reason, the study of AL did not really develop as a separate scientific study until rather recently when the computer had become a well-established tool in science. However, we must not forget that the study of AL has grown out of a long and fascinating history of ideas that predates the computer. AL is a science that has taped into certain basic drives in western culture that have sought to predict and control nature. Simon Penny suggests, in an essay printed in *Scientific American,* that the study of AL is motivated by a desire which has been deeply imbedded in western culture for centuries where:

"Artists and inventors have attempted to imitate nature (a process known as mimesis) and to simulate the qualities of being human (anthropomorphism). These twin drives, which lie at the very heart of western culture, blur the lines between animate and inanimate, between human and machine.

These cultural motivations first appear in the attempt to simulate life through artistic endeavors and leads eventually to the invention of mechanical simulations. As we will see these mechanical contrivances became more and more lifelike as our technology advanced. The anthropologist Stefan Helmreich has also noted the tendency for these deeply rooted drives to express themselves in the field of AL and has written persuasively on the subject. Helmreich sees AL as the current incarnation of a drive to create simulacra of living things that can trace its history through "...the Rabbi of Prague's Golem of clay, to the hydraulic and mechanical contraptions of mid-millennium, to Mary Shelly's Frankenstein, to the von Neumannian self-replicating automata....".

These cultural drives anthropomorphism and mimesis, transcend disciplinary boundaries and inhabit the most sophisticated technologies available at any historical moment. In the previous century there were many historical examples of mechanical simulations of life based on simple mechanical gears and pulleys animating a life-like puppet of some sort or another. These clockwork automatons, such as the famous mechanical ducks built by Jacques de Vaucanson in 1735, could eat, drink, quack, etc., in a way similar to a real duck. These automata where very compelling in their time but would not be considered an adequate scientific model of a truly living duck today by any but the most theory-blinded mechanist. Still Vaucanson was an early explorer of AL and as his modern counterparts do today he labored, "Under the influence of contemporary philosophic ideas, he tried, it seems, to reproduce life artificially".

Vaucanson was limited by the level of technology available to him in the creation of his models. His automata where typically mechanical puppets, were controlled by levers and pulleys which passed through the feet of the contrivance and lead to a much more complex piece of machinery bellow the stage. The unseen mechanisms below were responsible for the complex behavior of the puppet above. Still the puppet was not controlled by a human puppeteer, all of its motions were mechanically preprogrammed like clockwork and this was a truly novel innovation for the time. These contrivances where limited by the nature of their construction so that they could not, for instance, move very far and they where limited to a certain set of programmed motions. Any one seeing one of these mechanical ducks today would be impressed by the workmanship displayed in there construction (the wing alone was made of 400 parts), but no one would think of them as being alive or that these machines are a good model to study living ducks by.

In his article "Artificial Life," Christopher Langton sketches out the history of AL starting with the mechanical Ducks we have been discussing and going on to introduce other mechanical automatons. Although Langton does not state this explicitly, these automatons are the first attempts at making models that could be used to prove the theories of the philosophy of mechanism. But this pursuit did not produce anything more than drawing room curiosities until the modern era when the work of the great logicians such as "Church, Kleene, Gödel, Turing, and Post formalized the notion of a logical sequence of steps, leading to the realization that the essence of a mechanical process — the "thing" responsible for it dynamic behavior — is not a thing at all, but an abstract control structure, or "Program" — a sequence of simple actions selected from a finite repertoire". This conceptual freedom allowed theoreticians to move beyond the confines of simple *practical* mechanical processes and enter the heady realm of *theoretical* mechanical processes.

It is at this point in the history of ideas that modern AL studies truly begin. After the late 1940's we begin to see theoretical mathematicians and physicists taking an interest in answering questions concerning the phenomena of life. At this time we have Erwin Schodinger's book *"What Is Life?"* where he attempts to apply the mathematical rigor of physics to the study of biology, and we also see the beginnings of the mechanical theories of life developed by John von Neumann with his theory for self replicating automata. As von Neumann's automata are conceived of as a thought experiments and not as practically constructable machines it is clear that we have moved beyond the eccentric parlor games represented by the mechanical duck and into what we would recognize as modern serious science.

Von Neumann's automaton is a machine that is only conceivable as a logical process. The first version of his thought experiment has us imagine a machine floating around in a pond stocked with all the basic parts necessary to build an exact duplicate of the machine. Through a complicated logical process that von Neumann thoroughly describes, the machine is able to build an exact duplicate of itself, thus achieving self-replication. Von Neumann believed self-replication was the essential ingredient for a thing to be considered living. Von Neumann eventually abstracted this thought experiment even further as he felt, paradoxically, that it was necessary to remove all reference to reality in order to determine the essential logical structure of self reproduction. In order to achieve this abstraction from the real world von Neumann's final version of his self-replicating machine exists as a mathematical formula that is expressed through the mathematical tool of the cellular automaton.

Cellular Automata are a type of mathematical formula in which the variables are best conceived of as cells in a domain of some certain description. For example, we could make a two dimensional Cellular Automaton by drawing a checkerboard pattern on some paper and creating "rules" which will tell us what "State" each square is in during each time step of the program. For instance, each square could be blank or filled with a variable of some numerical value determined by the rules of the system. The rules used in determining the values of each "cell" in the automaton are usually something like this: if the cell is blank in this time step and three of its eight neighbors are filled then the cell is to be filled in the next time step.

These rules can be simple or complex depending on the needs of the programmer. Von Neumann's automaton is relatively complex, the cells in this simulation can be in any one of twenty-nine different states, when the cellular automaton's program is set in motion it can build an exact duplicate of itself using only the logical rules described in the program. Von Neumann was able to provide us with a purely formal description of mechanical self-reproduction. From that point on the cellular automaton has become the dominant paradigm used in the study of artificial life. Today there are many programs that use the cellular automaton to model living systems. Unfortunately the problems with cellular automata are that they are tedious, if not impossible, to work out by hand, and even though the behavior of a cellular automaton is completely deterministic, since the interactions between many cells is so complex there is no way to predict the behavior of the automaton without physically running through the entire simulation.

This unfortunate situation has a happy ending though. There is an easy technological fix to the weaknesses of the cellular automaton. Modern computers are extremely adept at running through the myriad of tedious calculations required to set the automaton in motion and, as an added benefit, the screen of a computer monitor can easily provide a compelling visual accompaniment to the cellular automaton as the pixels on the screen are perfect for representing the abstract cells in the automata, with different colors representing different variable states in the cells. Thus the technology of the computer has provided any interested AL researcher with the ability to acquire relatively cheep and speedy machines upon which he or she can then test out his or her theories. As one could easily guess, the most successful models of living systems have been created using the most advance computer technology available.

One might ask why so much faith is invested in computer technologies as the proper tool for studying life. Christopher Langton answers this question as follows:

> Computers should be viewed as *second-order* machines — given the formal specification of a first-order machine, they will "become" that machine. Thus, the space of possible machines is directly available for study, at the cost of a mere formal description: computers "realize" abstract machines.

Since life is preconceived of as a machine in mechanistic philosophies of life then it follows from the above that anyone with a good enough computer and a properly conceived program can make useful contributions to the life sciences.

So we can see that the study of AL is intimately tied to the current technology of computers. This means that the study of AL is completely determined by what the computer is currently able to do. The AL researcher not only has to be savvy in the fields of the life sciences he or she also has to be able to use computer technology to its fullest through a complete understanding of the hardware and the ability to program software.

The AL researcher has to be or have access to a good programmer. Through the skill of programming the AL researcher, in a very real way, crafts the world that he or she plans to study. This fact has put some theoretical biologists off from seeing AL as truly contributing to the knowledge of living systems and biology. The awkward situation of having a technology color a field of study so thoroughly places the science of AL squarely in the middle of a debate that has been going on for some time in the philosophy of technology. And that debate is over the question regarding what is the relationship between science and technology?

Science and Technology or Techno Science

The battle between the perceived superiority of science over technology is an old one dating back to the ancient Greeks or at least to the exhuming of Greek philosophical ideas in the middle ages. Technics have, to paint a broad picture, been seen to be the province of science applied, not science practiced. This is certainly true of the attitudes towards science and technology coined during the industrial revolution where, "...the now established nomenclature which relates the "pure" or "theoretical" sciences such as physics, to the "applied" sciences such as engineering were already in place.

This concept has been slowly breaking down under the pressure of how modern science is actually practiced with big money spent to pursue pragmatic problem solving goals. In the early phases of western history it might have been proper to argue that the scientific musings of the idle elite would have very little impact on the daily lives of the craftsmen and workers who kept the society operating, but today each new scientific discovery quickly suggests new technological applications and each new technology creates new tools to be used in scientific endeavors, high technology and science have become almost indistinguishable.

Early Modern Science was preceded by the beginnings of the world voyages and the discovery of the New World and the rise of science as distinct disciplines accompanied by the machine age of the Industrial Revolution, the twentieth century saw science become fully a *techno science* now thought to be the "motor" which drove and developed what is called Modern Technology, presumably distinct from any ancient or traditional technologies.

Without embarking on a lengthy detour we need to define more carefully what is meant by the term "techno science." This term was first used by Bruno Latour and it is used to refer to his belief that modern science is embodied in its instruments. What this means is that no longer is science practiced purely as an observation of natural events but that the entire experimental process is influenced entirely by the apparatus used to conduct the experiment, the scientist experiences the phenomena he or she studies *through* scientific instruments and without those instruments modern science is impossible, science and technology are now inseparable.

Artificial Life is the Techno Science of Biology

In contrast to other sciences like chemistry and physics, biology has long lagged behind in its ability to produce large-scale benefits to technology and society. There has long been a kind of physics envy in certain circles of the study of Biology. Traditionally Biology has not been the benefactor of big corporate or government funding and I would suggest that this is due to Biology's failure to define its ability to contribute to profitable technological advances in fields other than medicine. The belief that modern science should be seen as a techno science should come easy to any AL researcher. It is a common argument among A Lifers that the new science of AL has opened up an entirely new set of questions for biology to study.

No longer is theoretical Biology limited to questions concerning life-as-we-know-it, we can now, through the use of AL simulations on computers, look at questions derived from the study of life-as-it-could-be. Since, as we saw above, the science of AL was brought about through advances in computer technology, it is easy to assume that with each new advance in computer technology new techniques of studying AL will be enabled allowing more, as yet undreamed of, questions to be answered. AL's relationship with the technology of computers is reciprocal. The study of AL has already produced spin-off technologies that are beginning to see common uses in computer graphics and in programming techniques. An example of this is the growing use of the programming technique called the Genetic Algorithm Paradigm.

The Genetic Algorithm Paradigm is a relatively new form of computer programming that attempts to harness biologically inspired Darwinian evolutionary pressures in the creation of computer algorithms. This new programming technique was developed when AL researchers wanted to mimic the transfer of genes between two members of a species when matting occurs. In order to duplicate this phenomenon programming methods where developed where parts of the code that are used to described one *organism* could be "spliced" as it were to other parts of code which described a second *organism* thus creating a new hybrid computer program that described a third progeny *organism*. This was seen as an adequate model of sexual reproduction on a computer.

This model has since been modified and applied to more traditional problems in computer science and has been developed into a new technology that allows a computer to be set to a specific task and then starting from a group of random programs it "evolves" new programs by combining the "seed" programs in myriad's of combinations until one emerges that can solve the problem, thus the machine automatically programs itself. This is a truly novel invention and it stems from research in AL.

From this quick overview on can easily see that the science of AL is a perfect example of a techno science as it is described in the literature of the philosophy of technology. The questions that the science of AL attempts to answer are formed by the technology used to answer them, and in turn the technology used by the science is modified by the demands and breakthroughs of the science. Since AL is an outgrowth of the study of theoretical biology the useful additions to computer technology should allow theoretical biology to begin to move into the powerful interests of techno science which previously has been the exclusive realm of physics and chemistry.

Some Thoughts on Techno Science

The concept of a techno science is unsettling to those who believe that science has a privileged access to the demonstrable truths of reality. Once we begin to accept that science can mix with, and in some ways be determined by another human praxis like technology makes it harder to argue that science is not just one human praxis among many but have the ability to make adequate statements about reality. This is the kind of argument that frightens commentators on science such as Horgan, because these arguments would seem to open the door to uncontrolled relativism and a resulting epistemological paralysis.

This is another point in which a lengthy digression could follow but instead I will just mention that this concern has been quite effectively ameliorated by feminist theories of science such as Sandra Harding's *Whose Science? Whose Knowledge?* In her view, the realization that traditional science is not an automatically privileged view of the truth is a liberating experience that allows us to pursue a new science that will be "less partial and less distorted," and therefore more true to the actual goals of the pursuit of knowledge.

The belief that a progressive positivistic science and technology will somehow automatically save us from all the ills of our times is seen to be laughably naive today. It is of no use, as Horgan is want to do, to pine for the lost days of the heroic age of science, they never really existed anyway. It is much more useful in a pragmatic sense to acknowledge that science is now (and probably always was), dependent on technology and that science is a human product just like technology and does not posses the strange mystical ability to discover disembodied transcendent truths. Understanding the weaknesses of a synthetic science is an important step in making that science work to our best interests. Seeing science as techno science allows us to create a clearer conception of science as compared to technology and their proper relation to each other.

The systematic and stable interconnection of scientific and technological practices and institutions into a "techno science" is usually considered as the outcome of developments in the twentieth century, with forerunners in the second half of the nineteenth century. In the time period before the late nineteenth century, according to the common view, there may have been exchanges of knowledge, skill, and instruments between savants experimenting at academic institutions and craftsmen and artisans producing in workshops, but this interaction did not result in an intricate entanglement of these two different cultures. Chemistry prior to the second half of the nineteenth century and the emergence of the synthetic-dye industry apparently fits this general picture. Although it has been acknowledged that chemical technology occasionally stimulated developments in eighteenth-century academic chemistry, and vice versa, chemical science and technology have largely been studied historically as two separate domains.

By contrast, it is the central thesis of this project, sustained by previous historical studies, that long before the second half of the nineteenth century, chemical science and technology were strongly and systematically linked with each other. The project aims to unravel these links, with a focus on the eighteenth century. The interconnection of chemical science and technology was not merely promoted by a few individuals occupied with both scientific and technological enterprises. Rather, this interconnection was firmly established on the communal and institutional level of eighteenth-century chemistry. It was entrenched in a shared material culture that spanned from the academic laboratory to the chemical workshop, comprising material objects of inquiry, instruments, reagents, types of manipulations and the site of experimentation.

Comparable to the systematic and permanent alliances between science and technology in the twentieth century, eighteenth-century chemistry was an early form of techno science. Unlike the later techno scientific alliances, however, this early form was not the result of a convergence of previously separated scientific and technological cultures and practices, but rather a hybrid scientific-technological endeavor from the very beginning of its institutionalization in late seventeenth-century Europe. After the first successful steps were taken in the seventeenth century to institutionalize chemistry in academies, medical faculties, botanical gardens, and museums, in the eighteenth century chemistry was an established part of the intellectual world. Eighteenth-century chemists were teachers and professors, authors of learned books and experimental reports, members of academies and scientific societies, and visitors of coffee shops and salons.

Yet, they differed markedly from other savants of the time; not only because they spent many hours of the day experimenting in the laboratory, but also because of their diverse technological and commercial activities. Among the eighteenth-century European chemists were many apothecaries, metallurgical officials and consultants, inspectors of manufactories, members of state committees and technological boards, and entrepreneurs. Eighteenth-century chemists working at medical faculties, professional schools and other academic institutions instructed their students about pharmaceutical techniques and various areas of practical, artisanal chemistry. In their laboratories they repeated artisanal operations and analyzed materials produced and applied in the chemical arts and crafts, using almost exclusively, in the first half of the eighteenth century, instruments shared with assayers, apothecaries and other artisans.

Eighteenth-century chemists not only shared a few single instruments with apothecaries and other artisans, but also made use of an arsenal of instruments used in the arts and crafts. Their smelting and testing furnaces, bellows, crucibles, calcinations dishes, and balances were similar to the instruments used by assayers. The same types of mortars, pestles, filters, vessels, boxes, glass tubes, vials, retorts, alembics, pelicans, receivers, and transmission vessels that academic chemists used in their laboratories were also used by distillers for producing nitric acid, alcoholic spirits and fragrant oils, and by apothecaries for making medicines.

There was even agreement on the size of vessels and instruments used by academic chemists, apothecaries, assayers and distillers. The small-scale trial was intrinsic to assaying, which studied the composition of ores and other minerals for calculating the productiveness of mining and metallurgy. As pharmacy was still a handicraft in the eighteenth century, it also produced remedies on a small scale and for a comparatively small community of local consumers. The distilling of essential oils for the making of perfumes and alcoholic spirits was performed on a small scale too, even though there was enormous modification of distilling apparatus used by commercial distillers.

Unlike the core areas of eighteenth-century experimental philosophy, which became transformed into "experimental physics" in the early nineteenth century, chemistry was a culture in which experimentation took place on a more or less everyday basis at a specific workplace, the "laboratorium." In the eighteenth century, the term "laboratory" referred mostly to workplaces where chemical and pharmaceutical operations were performed, and sometimes to those of assayers, distillers, and manufacturers of gunpowder, too. The Latin word laborare, from which laboratory is derived points to its similarity with ordinary workshops. By contrast, in the core areas of experimental philosophy, "physical cabinets" and "physical theaters" (theatrum physicum) were established.

As is more or less apparent from these latter terms, the institutions frequently served as locations for the collection and exhibition of instruments and for the demonstration of curious experimental effects, rather than as actual places of work. The resemblance of the material equipment of eighteenth-century chemical laboratories to that used at certain artisanal workplaces accorded with the fact that most chemists' types of experimental techniques corresponded with artisanal operations, in particular those of apothecaries, assayers and distillers. Dissolution, distillation, evaporation, precipitation, combustion and smelting were types of operations performed by both chemists and artisans. As many of the material substances studied in academic chemists' laboratories were bought from merchants and applied in practice by apothecaries and other artisans, the correspondence between the material culture of chemical science and chemical technology was quite strong.

The power of science was the epistemological strategy that powered the Enlightenment project. Scientific discovery and industrial technology underwrote modernity as these technologies reorganized the social and economic landscapes. The principles of science took form in the cult of Positivism, and the belief that scientific methodology could be applied to social organization that once the laws of human behavior were laid out the good society based on those laws could be constructed. The scientific paradigm slowly but surely displace religion as the Western worldview. And while the horrors of industrial capitalism impassioned the writings of Karl Marx and Charles Dickens, the achievements of industrial capitalism inspired reformers and writers like Edward Bellamy.

As Marx observed, the bourgeoisie released more industrial power in a hundred years than humankind in all the previous centuries. This power of transformation made possible by scientific discovery and application is a primary tenet of all modern narratives. But then the application of science seems to have gone awry. Huge military machines moved across the globe in two world wars culminating in Hiroshima. But even then America arose as a superpower proclaiming democracy and economic success for all. And while the US put a man on the moon it also fell prey to its own mythology in Vietnam. With the series of economic, political, and cultural crises that followed the Enlightenment project that had become intertwined with American appeared to have collapsed. The Western grand narrative that began in Europe seem to die in the colony it spawned.

Throughout the 20th Century, the United States has been identified with technology, and it dominated the world, in some large measure, because it ruled the technological roost. Especially in military technologies, the US consistently maintained the upper hand. And, beginning early in the century Capital and its representatives began associating Capital with advances in technological control and mastery. By mid century, GE's slogan, "Progress is Our Most Important Product," had gained hegemonic status. Though the slogans have changed, the gist — and that's really all there is now — of the message remains largely unchanged circa 1998. Ideologically, many of the ads under observation continue to push what Jurgen Habermas (1970, 81-122) described as civil privatism where a system of depoliticized, technocratic rule draws its legitimacy from providing rewards geared to privatized needs. Stuart Ewen has charted the transformation in American life from the citizen to the consumer.

Beginning in the 1920s, but greatly accelerated and intensified in the postwar 1950s, material abundance was cast as a product of highly efficient automated technologies "brought to you" by corporations such as GE, RCA, and GM. Disneyland's, the original Disneyland — and not any of the bigger, better, newer Disney thing parks — epitomized mass media accounts of how science and technology made for a more satisfying personal existence and a better society to boot. Walt Disney loved technology, just loved it and he built "Tomorrowland" as an homage to the promise of a utopian future built around the intelligent discoveries of Science applied to our daily lives in the form of commercial products. Even then in the late 1950s, Disney caught a stark, cold glimpse of a future mediated by the commodities developed and delivered by corporations — he captured a sense of the privatism of the family as a consuming unit, cut off visually from any wider community or society, encased in diorama and modern science. Sleek and clean, insulated (cut off) from all surrounding life — engaged in leisure moments — this was the utopian vision of the future.

In the 1970's Disney's vision took new form with the building of Epcot Center. Here, corporate sponsorship of the Disney pavilions collapsed entertainment, education, and public relations. The pavilions were part museum, part arcade, part theater, and part corporate show room. Each celebrated the achievements of applied science as indicators of human progress. Many commercials in our data set use the same signifiers found in Disney pavilions, such as light beams, robotic labor, and voice-over narratives of progress. These commercials mimic Disney's myth of progress by constructing elements of a technotopia without any underlying contradictions associated with techno science.

Empty Piers

Longshoremen on the Guaranteed Annual Income (GAI) are paid even after their work is technologically redundant within the shipping industry because they are guaranteed work or income when containerization of cargo eliminates their specific jobs. In effect since 1966, the GAI assures a full year's salary (2080 hours) at $45,000 a year to workers. The Union's premise in gaining this concession was that the cost of technological changes should not entirely rest upon the employees in the industry. In the 1950s the New York-New Jersey docks were being worked by 48,000 men who worked a total of 46,000,000 man hours and moved a total of 22,000,000 tons of general cargo. Their work was concentrated on the Manhattan and Brooklyn docks where Italian and black longshoremen moved cargo on the Brooklyn waterfront and Irish and Slavic longshoremen worked the Manhattan docks.

By the mid-70s, 12,000 longshoremen were working 22,000,000 man hours and moving 27,000,000 tons of cargo. In short, one-fourth the men moved more cargo tonnage. Longshoremen had increased their productivity by 400% in twenty years. With the increased "productivity" of the past two decades there has been a continuous decline in long shore job opportunities. The work has moved out of New York City to the modern container facilities in New Jersey, but no new job creation is occurring in New Jersey. In 1985, 10,100 longshoremen still worked the port of New York-New Jersey. Three thousand of these men on the GAI with high seniority who have been displaced by container technology rarely work but they receive a full salary. Even with the GAI this is a very profitable industry, generating $14 billion in economic activity and $2.3 billion in business income.

Faced with the technological elimination of their jobs, Brooklyn longshoremen attached themselves to a non productivist ideology. They collectively struggled for wages without work and bargained for free time as opposed to work time. No longer is their well-being attached to hard work as it was in the past; now it is attached to managing their own time away from work. They perceive that productivity in the workplace is opposed to their interests. The economists who argue the opposite do not represent them and they know it. Productivity means increased profits for the shipping industry and the elimination of the livelihood of the longshoremen.

The longshoremen's struggle is one for repossession of their time. Time has always been a contested terrain in the clash between labor and capital. In the past, the assumption that a hard day's work was both a necessity and a moral obligation undermined efforts to reduce the workday. In the workplace of computer-mediated production, robotics and information systems —workers must struggle to make both technology and time negotiable issues. But the longshoremen's struggle was shortsighted because they only struggled for their own immediate needs. They didn't understand that universal featherbedding is the only way to guarantee jobs at the wage and benefit levels they had achieved. They didn't struggle for a guaranteed permanent workforce in the port of New York-New Jersey, nor for work-sharing, nor for a shorter work day without a decrease in pay, and thus long shoring is an increasingly redundant occupation.

Now more and more of the work is by machine. The machines do everything: rolling, piping, felling. They put a gadget on the machine, like a funnel and it rolls and stitches at the same time. Now piping, we used to call it French piping, is done by machines. We used to do bottoms by hand and now the machines do the felling. We used to do everything by hand, but now it's like an assembly line and in the old days it was all custom made. We did everything by hand, now everything is by machine. They call it progress. They used to have twenty people do the work, all by hand. Now they have three people doing the same work and all by machine.

The only new work they create is in the sweet shops in Chinatown, Greenpoint, and Brighton Beach. These people work for peanuts. They can't collect unemployment, they have no benefits, and they work harder than we did. People need to make a living wage, people got to live. But it's always the same baloney. The immigrants are afraid, and everybody else wears blinkers."

These stories provide an overview of technological trends that have been evolving and now are occurring at an increasingly rapid pace. They affect the workforce from the manual and skilled laborer to the scientific researcher. Not only do they change the traditional world of work and challenge definitions of productivity, but they tear at the social fabric of the nation. The experiences contained in these stories suggest the following two conclusions for workers and for unions which traditionally have safeguarded the rights of employees:

1. In general, new high-tech production regimes are labor destroying. As a result, increasing numbers of workers are becoming permanently redundant. The wave of the future is more workers in a global labor market with fewer opportunities. Workers "of all collars" are increasingly being forced into greater competition with each other. In light of this situation, the initial struggle of longshoremen for the Guaranteed Annual Income has to become the struggle of all workers in all labor processes: manual, skilled, and intellectual. That is the struggle for basic income, independent of work.
2. Technological and scientific knowledge has become the principal productive force in late industrial societies. Not only has manual work been displaced but skilled work has been displaced as well. In the new workplaces techno science is dominant and skilled work is moved to the margins of production. Both skilled and industrial unionism faces an increasing dilemma in view of these developments. If unions are to survive, they must

reconceptualize their organizing strategies, with an increasing emphasis on knowledge work. This is already beginning to occur.

With more and more workers now faced with the technological elimination of their work, and the lack of emerging occupations to sustain a good life for all, two challenges and responses emerge. First, the work ethic is no longer a central organizing principle for social life, and second, there is a general decline in wages. This suggests two responses. We must actively participate in the creation of a *new public ethic of social responsibility* around which life now can be organized. Second, *the struggle for the decommodification of basic needs:* medicine, housing, education, food, etc. has to be initiated.

In this age of new social movements, Foucault is right that micro struggles are everywhere. From communitarianism to new socialists, from the new right to pro-choice, micro-politics is everywhere. None of these struggles offers an alternative to the present situation because each is focused against the consequences of a problem and never against the premises of the political, economic, cultural order. They have not created an alternative discourse that can nourish a movement to reconstruct the world. Micro-struggles are insufficient for the task. They may help to define a new situation, even provide some innovations, but their language is pluralist, never offering anything more than temporary reform. A world which can no longer be organized around the work ethic requires a new organizing principle—an alternative discourse. A new public ethic of social responsibility is an attempt at constructing this alternative discourse.

This new public ethic of social responsibility does not ask for a "return to family values" or identify the cause of social problems as "compassion deficit." It doesn't speak of social problems because it doesn't speak the language of consequences. This ethic of social responsibility speaks the language of change which has as its first premise understanding the enormous human and environmental costs of disorganized capitalism and an unrestrained global free market. An ethic of social responsibility assumes that market based solutions which reward some are much too expensive for the great majority of the world's population. These costs include the perpetuation of all forms of class, race and gender oppression.

The "jobless future" negates any possibility that the capitalist marketplace can end the current misery of many in the world. The International Labor Organization measured worldwide unemployment and underemployment in 1994 at 820 million and growing. It also states that "in total some 1.1 billion people around the world live below the poverty level of their own country because of a lack of a good paying occupation."

The "jobless future" requires an alternative to a world based on work and jobs. Thus the proposed public ethic of social responsibility understands that there is work to be done but it has been deemed too costly under capitalism. Thus schools aren't built, housing isn't built, health care isn't guaranteed for everyone, the air isn't made clean, people aren't fed even though we already have the capabilities of accomplishing these human necessities. Only in a world where persons come before economics can we rebuild and guarantee a right to a decent life to all.

To achieve a worldwide ethic of social responsibility will require struggle in the workplace. A thirty-hour week and a six-hour day with no overtime and no decrease in income is important. This is required to create new opportunities and for the health of the worker. Survival for too many today requires overtime or "two-job existence" and this means a reduction in the standard of living. Decommodification—removing health care, housing, education, food from the marketplace would enable all to have the "good life" not as a privilege but as a right. Decommodification then is not just removing central human necessities from the marketplace but, more importantly, it is the removal of humans as workers from their own status as commodities.

Decommodification as a central component of the new ethic of social responsibility is the struggle for a world in which people's lives are no longer dominated by jobs and trying to "make ends meet." Free of this they can now be citizens who can participate in their world. Politics requires free time, once only available to the few, but ideally made possible to everyone. Citizenship should not be centered on voting alone but on participation and governance. For this to occur women and men must struggle for a world in which there is freedom from work and want and freedom for human life.

8

Nuclear Technology, Radioactive Waste and Human Value

Radioactivity arises naturally from the decay of particular forms of some elements, called isotopes. Some isotopes are radioactive, most are not, though in this publication we concentrate on the former. There are three kinds of radiation to consider: alpha, beta and gamma. A fourth kind, neutron radiation, generally only occurs inside a nuclear reactor.

Different types of radiation require different forms of protection:

- Alpha radiation cannot penetrate the skin and can be blocked out by a sheet of paper, but is dangerous in the lung.
- Beta radiation can penetrate into the body but can be blocked out by a sheet of aluminium foil.
- Gamma radiation can go right through the body and requires several centimeters of lead or concrete, or a meter or so of water, to block it.

All of these kinds of radiation are, at low levels, naturally part of our environment. Any or all of them may be present in any classification of waste.

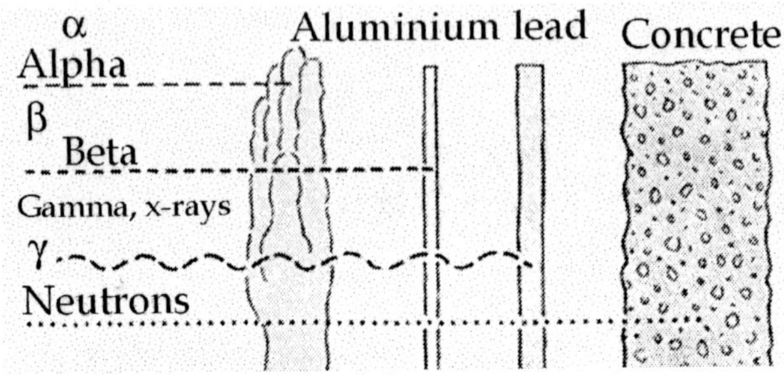

Fig: 8.1

Radioactive wastes comprise a variety of materials requiring different types of management to protect people and the environment. They are normally classified as low-level, medium-level or high-level wastes, according to the amount and types of radioactivity in them.

Another factor in managing wastes is the time that they are likely to remain hazardous. This depends on the kinds of radioactive isotopes in them, and particularly the half lives characteristic of each of those isotopes. The half life is the time it takes for a given radioactive isotope to lose half of its radioactivity. After four half lives the level of radioactivity is 1/16th of the original and after eight half lives 1/256th. The various radioactive isotopes have half lives ranging from fractions of a second to minutes, hours or days, through to billions of years. Radioactivity decreases with time as these isotopes decay into stable, non-radioactive ones.

The rate of decay of an isotope is inversely proportional to its half life; a short half life means that it decays rapidly. Hence, for each kind of radiation, the higher the intensity of radioactivity in a given amount of material, the shorter the half lives involved.

Three general principles are employed in the management of radioactive wastes:

- Concentrate-and-contain.
- Dilute-and-disperse.
- Delay-and-decay.

The first two are also used in the management of non-radioactive wastes. The waste is either concentrated or then isolated, or it is diluted to acceptable levels and then discharged to the environment. Delay-and-decay however is unique to radioactive waste management; it means that the waste is stored and its radioactivity is allowed to decrease naturally through decay of the radioisotopes in it.

Types of Radioactive Waste

Low-level Waste is generated from hospitals, laboratories and industry, as well as the nuclear fuel cycle. It comprises paper, rags, tools, clothing, and filters, etc. which contain small amounts of mostly short-lived radioactivity. It is not dangerous to handle, but must be disposed of more carefully than normal garbage. Usually it is buried in shallow landfill sites. To reduce its volume, it is often compacted or incinerated (in a closed container) before disposal. Worldwide it comprises 90% of the volume but only 1% of the radioactivity of all radwaste. Intermediate-level Waste contains higher amounts of radioactivity and may require special shielding.

It typically comprises resins, chemical sludges and reactor components, as well as contaminated materials from reactor decommissioning. Worldwide it makes up 7% of the volume and has 4% of the radioactivity of all radwaste. It may be solidified in concrete or bitumen for disposal. Generally short-lived waste (mainly from reactors) is buried, but long-lived waste (from reprocessing nuclear fuel) will be disposed of deep underground. High-level Waste may be the used fuel itself, or the principal waste from reprocessing this. While only 3% of the volume of all radwaste, it holds 95% of the radioactivity. It contains the highly-radioactive fission products and some heavy elements with long-lived radioactivity.

It generates a considerable amount of heat and requires cooling, as well as special shielding during handling and transport. If the used fuel is reprocessed, the separated waste is vitrified by incorporating it into borosilicate (Pyrex) glass which is sealed inside stainless steel canisters for eventual disposal deep underground. On the other hand, if used reactor fuel is not reprocessed, all the highly-radioactive isotopes remain in it, and so the whole fuel assemblies are treated as high-level waste. This used fuel takes up about nine times the volume of equivalent vitrified high-level waste which results from reprocessing and which is encapsulated ready for disposal.

Both high-level waste and used fuel are very radioactive and people handling them must be shielded from their radiation. Such materials are shipped in special containers which prevent the radiation leaking out and which will not rupture in an accident. Whether used fuel is reprocessed or not, the volume of high-level waste is modest, - about 3 cubic meters per year of vitrified waste or 25-30 tonnes of used fuel for a typical large nuclear reactor. The relatively small amount involved allows it to be effectively and economically isolated.

Radioactive Materials in the Natural Environment

Naturally-occurring radioactive materials are widespread throughout the environment, although concentrations are very low and they are not normally harmful. Soil naturally contains a variety of radioactive materials - uranium, thorium, radium and the radioactive gas radon which is continually escaping to the atmosphere. Many parts of the earth's crust are more radioactive than the low-level waste described above. Radiation is not something which arises just from using uranium to produce electricity, although the mining and milling of uranium and some other ores brings these radioactive materials into closer contact with people, and in the case of radon and its daughter products, speeds up their release to the atmosphere.

Wastes from the Nuclear Fuel Cycle

Radioactive wastes occur at all stages of the nuclear fuel cycle - the process of producing electricity from nuclear materials. The fuel cycle comprises the mining and milling of the uranium ore, its processing and fabrication into nuclear fuel, its use in the reactor, the treatment of the used fuel taken from the reactor after use and finally, disposal of the wastes. The fuel cycle is often considered as two parts - the "front end" which stretches from mining through to the use of uranium in the reactor - and the "back end" which covers the removal of used fuel from the reactor and its subsequent treatment and disposal. This is where radioactive wastes are a major issue.

Residual Materials from the "Front End" of the Fuel Cycle

The annual fuel requirement for a 1000 MWe light water reactor is about 25 tonnes of enriched uranium oxide. This requires the mining and milling of perhaps 50,000 tonnes of ore to provide about 200 tonnes of uranium oxide concentrate (U_3O_8) from the mine. At uranium mines, dust is controlled to minimise inhalation of radioactive minerals, while radon gas concentrations are kept to a minimum by good ventilation and dispersion in large volumes of air. At the mill, dust is collected and fed back into the process, while radon gas is diluted and dispersed to the atmosphere in large volumes of air.

At the mine, residual ground rock from the milling operation contains most of the radioactive materials from the ore, such as radium. This material is discharged into tailings dams which retain the remaining solids and prevent any seepage of the liquid. The tailings contain about 70% of the radioactivity in the original ore.

Eventually these tailings may be put back into the mine or they may be covered with rock and clay, and then revegetated. In this case considerable care is taken to ensure their long-term stability and to avoid any environmental impact (which would be more from acid leaching or dust than from radioactivity as such). The tailings are usually around ten times more radioactive than typical granites, such as used on city buildings. If someone were to live continuously on top of the Ranger tailings they would receive about double their normal radiation dose from the actual tailings (i.e. they would triple their received dose).

With in situ leach (ISL) mining, dissolved materials other than uranium are simply returned underground from where they came, as the water is recirculated. Uranium oxide (U_3O_8) produced from the mining and milling of uranium ore is only mildly radioactive - most of the radioactivity in the original ore remains at the mine site in the tailings. Turning uranium oxide concentrate into a useable fuel has no effect on levels of radioactivity and does not produce significant waste. First, the uranium oxide is converted into a gas, uranium hexafluoride (UF_6), as feedstock for the enrichment process.

Then, during enrichment, every tonne of uranium hexafluoride becomes separated into about 130 kg of enriched UF_6 (about 3.5% U-235) and 870 kg of 'depleted' UF_6 (mostly U-238). The enriched UF_6 is finally converted into uranium dioxide (UO_2) powder and pressed into fuel pellets which are encased in zirconium alloy tubes to form fuel rods. Depleted uranium has few uses, though with a high density (specific gravity of 18.7) it has found uses in the keels of yachts, aircraft control surface counterweights, anti-tank ammunition and radiation shielding. It is also a potential energy source for particular (fast neutron) reactors.

Wastes from the "Back End" of the Fuel Cycle

It is when uranium is used in the reactor that significant quantities of highly radioactive wastes are created. When the uranium-235 atom is split it forms fission products, which are very radioactive and make up the main portion of nuclear wastes retained in the fuel rods. There is also a relatively small amount of radioactivity induced in the reactor components by neutron irradiation. About 25 tonnes of used fuel is taken each year from the core of an 1000 MWe nuclear reactor. This fuel can be regarded entirely as waste (as, for 40% of the world's output, in USA and Canada), or it can be reprocessed (as in Europe and Japan). Whichever option is chosen, the used fuel is first stored for several years under water in cooling ponds at the reactor site. The concrete ponds and the water covering the fuel assemblies provide radiation protection, while removing the heat generated during radioactive decay.

The costs of dealing with this high-level waste are built into electricity tariffs. For instance, in the USA, consumers pay 0.1 cents per kilowatt-hour, which utilities pay into a special fund. So far more than US$ 18 billion has been collected thus.

Reprocessing

If the used fuel is later reprocessed, it is dissolved and separated chemically into uranium, plutonium and high-level waste solutions. About 97% of the used fuel can be recycled leaving only 3% as high-level waste. The recyclable portion is mostly uranium depleted to less than 1% U-235, with some plutonium, which is most valuable.

Arising from a year's operation of a typical 1000 MWe nuclear reactor, about 230 kilograms of plutonium (1% of the spent fuel) is separated in reprocessing. This can be used in fresh mixed oxide (MOX) fuel (but not weapons, due its composition). MOX fuel fabrication occurs in Europe, with some 25 years of operating experience. The main plant is in France, and started up in 1995. Japan's slightly smaller plant is due to start up in 2012. Across Europe, over 35 reactors are licensed to load 20-50% of their cores with MOX fuel.

The separated high-level wastes - about 3% of the typical reactor's used fuel - amounts to 700 kg per year and it needs to be isolated from the environment for a very long time.

Major commercial reprocessing plants are operating in France and UK, with capacity of almost 5000 tonnes of spent fuel per year, - equivalent to at least one third of the world's annual output. A total of over 90,000 tonnes of spent fuel has been reprocessed at these over 40 years.

Solidification processes have been developed in several countries over the past fifty years. Liquid high-level wastes are evaporated to solids, mixed with glass-forming materials, melted and poured into robust stainless steel canisters which are then sealed by welding. The vitrified waste from the operation of a 1000 MWe reactor for one year would fill about twelve canisters, each 1.3m high and 0.4m diameter and holding 400 kg of glass. Commercial vitrification plants in Europe produce about 1000 tonnes per year of such vitrified waste (2500 canisters) and some have been operating for more than 20 years.

A more sophisticated method of immobilising high-level radioactive wastes has been developed in Australia. Called 'SYNROC' (synthetic rock), the radioactive wastes are incorporated in the crystal lattices of the naturally-stable minerals in a synthetic rock. In other words, copying what happens in nature. This process is now being tested in USA.

Waste Disposal

Final disposal of high-level waste is delayed for 40-50 years to allow its radioactivity to decay, after which less than one thousandth of its initial radioactivity remains, and it is much easier to handle. Hence canisters of vitrified waste, or used fuel assemblies, are stored under water in special ponds, or in dry concrete structures or casks for at least this length of time. The ultimate disposal of vitrified wastes, or of used fuel assemblies without reprocessing, requires their isolation from the environment for long periods. The most favoured method is burial in dry, stable geological formations some 500 meters deep. Several countries are investigating sites that would be technically and publicly acceptable. The USA is pushing ahead with a repository site in Nevada for the entire nation's used fuel.

One purpose-built deep geological repository for long-lived nuclear waste (though only from defence applications) is already operating in New Mexico. After being buried for about 1000 years most of the radioactivity will have decayed. The amount of radioactivity then remaining would be similar to that of the naturally-occurring uranium ore from which it originated, though it would be more concentrated.

Layers of Protection

To ensure that no significant environmental releases occur over a long period after disposal, a 'multiple barrier' disposal concept is used to immobilise the radioactive elements in high-level (and some intermediate-level) wastes and to isolate them from the biosphere. The principal barriers are:

- Immobilise waste in an insoluble matrix, e.g. borosilicate glass, Synroc (or leave them as uranium oxide fuel pellets - a ceramic).
- Seal inside a corrosion-resistant container, e.g. stainless steel.
- Surround containers with bentonite clay to inhibit any groundwater movement if the repository is likely to be wet.
- Locate deep underground in a stable rock structure.

For any of the radioactivity to reach human populations or the environment, all of these barriers would need to be breached before the radioactivity decayed. In USA high-level civil wastes all remain as used fuel stored at the reactor sites. It is planned to encapsulate these fuel assemblies and dispose of them in an underground engineered repository at Yucca Mountain, Nevada. This is the program which has been funded by electricity consumers to US$ 26 billion so far (i.e. @ 0.1 cent per kWh), of which about US$ 6 billion has been spent.

In Europe some used fuel is stored at reactor sites, similarly awaiting disposal. However, much of the European spent fuel is sent for reprocessing at either Sellafield in UK or La Hague in France. The recovered uranium and plutonium is then returned to the owners (the plutonium via a MOX fuel fabrication plant) and the separated wastes (about 3% of the spent fuel) are vitrified, sealed into stainless steel canisters, and either stored or returned. Eventually they too will go to geological disposal. Sweden represents the main difference. It has centralised used fuel storage at CLAB near Oskarshamn, and will encapsulate used fuel there for geological disposal by about 2015. Finland is establishing a final repository at Olkiluoto. European funding is at similar level to USA per kWh.

A Natural Precedent

We have an example in nature to suggest that final disposal of high-level wastes underground is safe. Two billion years ago at Oklo in Gabon, West Africa, and chain reactions started spontaneously in concentrated deposits of uranium ore. These natural nuclear reactors continued operating for hundreds of thousands of years forming plutonium and all the highly radioactive waste products created today in a nuclear power reactor. Despite the existence at the time of large quantities of water in the area, these materials stayed where they were formed and eventually decayed into non-radioactive elements. The evidence is there.

Alternatives to Nuclear Electricity

No technology is absolutely safe or without environmental effects. We should therefore compare the production of electricity from nuclear energy with the other options available to us. Burning coal in power stations is still the major source of electricity worldwide, followed by hydro, uranium and gas. A 1000 MWe light water reactor uses about 25 tonnes of enriched uranium a year, requiring the mining of some 50,000 tonnes of uranium ore. By comparison, a 1000 MWe coal-fired power station requires the mining, transportation, storage and burning of about 3.2 million tonnes of black coal per year.

This creates around 7 million tonnes of carbon dioxide not to mention sulfur dioxide, depending on the particular coal. Solid wastes from a coal-fired power station can be substantial and cause environmental and health damage. Many people are concerned about possible global warming through enhancement of the greenhouse effect. Much of this is due to steadily increasing carbon dioxide in the atmosphere over the past 150 years. Burning fossil fuels, particularly coal, for electricity contributes almost 10 billion tonnes of carbon dioxide to the air each year.

To investigate or consider:

- Why are nuclear wastes sometimes said to be a problem which is too difficult to solve?
- What are the advantages and disadvantages of the two ways of dealing with high-level waste (reprocessing and vitrification, or treating whole fuel assemblies as waste)?
- How do nuclear wastes compare with other industrial wastes? (Look at their hazard, the care which is taken with them and the funding involved.)
- What other industrial wastes decay over time so that their hazard steadily diminishes?
- How are the wastes from coal-fired electricity generation disposed of?

The Nuclear Waste

To some extent all uses of energy affect the environment adversely. But over the centuries, waterwheels grinding grain, windmills pumping water, and animals ploughing land did little environmental damage. The damage really began in the seventeenth century with the invention of the steam engine and mining coal to fuel it. Growing volumes of waste effluent from burning coal, and later the other fossil fuels, oil and gas, have brought us to the environmental crisis facing us today.

Now in the twentieth century forests die from acid rain, marine life is blighted by oil spills and urban populations suffer health problems from the smog created by commuter traffic. Cleaner burning of fossil fuels could do a lot to reduce the pollution. However, we now discover that the seemingly innocuous carbon dioxide from burning vast quantities of fossil fuels has been warming our globe since the industrial revolution though far from the only culprit carbon dioxide is the major offender.

Uranium is a fuel that releases no carbon dioxide and so it seems a promising candidate to displace fossil fuels at least to generate electricity. That nuclear power has serious problems is now generally acknowledged even by its advocates. However, climate change can be expected to bring great social disruption; so why not solve the problems of an energy source which emits no carbon dioxide.

The nuclear problem arousing most public concern is the disposal of radioactive waste. Wherever waste disposal sites are chosen local opposition has forced the plans to be cancelled. Radioactive wastes are produced at all stages of the nuclear fuel cycle. The wastes are stored as liquids or solids. When considering the nature of nuclear wastes it is helpful to see the nuclear fuel cycle as having a 'front end" production of the fuel for the nuclear reactor and a 'back end' storing and reprocessing the spent-fuel.

The 'Front End'

Mining

Waste discharge into the environment begins with mining uranium. Although the radioactivity of the wastes is low-level the volumes left behind at abandoned mining sites are huge. Only about three kilograms of uranium oxide are recovered from each tonne of ore. The left-over, looking like liquid mud and called tailings, are pumped into a tailings dam.

After mining ceases tailings dams become hills of fine sand which retain 80 per cent of the radioactivity in the ore body? Thorium-230 in the tailings decays into radium-226 which in turn decays into gaseous radon. Before mining, these and other radioactive elements are locked into impervious rock; little of it reaches the open environment and once mined they get into waterways and the atmosphere. Tailings are so finely ground that the radon escapes 10,000 times faster than from the ore body.

Tailings dams have a poor track record. Waterways have been polluted by radium after a sudden collapse or constant erosion. Radon gas and radioactive dust are carried downwind. Since radioactive decay will persist for over 100,000 years the hazard will be virtually without end. Most of the world's uranium has been mined in the poor African countries of Namibia and Niger and on the remnants of land left to the indigenous people of North America. The mining sites have become radioactive wastelands bringing devastating health problems to indigenous peoples.

Until the mid-1980s Colorado Plateau in the south-west of the United States provided half the world's uranium. The mining on the plateau is in the dry lands of the Navajo people who were promised jobs and wealth if they accepted mining. Today their lands are poisoned by dry dusty hills of tailings; winds spread radioactive dust and radon gas gets into the atmosphere. Aquatic creatures concentrate radium in the waterways up to 10,000 times. Navajo communities fear the consequences of eating the food they grow because it has elevated levels of radioactivity. They suffer an epidemic of cancer and birth defects.

A report by the Los Alamos laboratory, a leading centre of nuclear research, says the health problems created by uranium mining may be irremediable. The report says the only solution could be "to zone the land in uranium mining and milling districts to forbid human habitation". Australia has wastelands of its own, a legacy of mining uranium in the past. They are a portent of what could happen at Ranger and Roxby Downs after mining ceases. Uranium ore mined at Radium Hill in South Australia was once processed into yellowcake at Port Pirie situated on the Spencers Gulf. The radioactive tailings from the operation were left to be washed away by high tides and eroded by high winds.

At Mary Kathleen in Queensland tailings, which were poorly rehabilitated with a crushed rock cover, now seeps radioactive waste into underground waters. In the Northern Territory monsoonal rains breached tailings dams of abandoned mines at Northern Hercules and Rum Jungle poisoning long stretches of the rivers below them. Ranger mine is situated in the Kakadu National Park listed as a World Heritage. Australia is committed to protect the park's natural wonders. The mine's tailings dam, situated in the path of monsoonal floods that pour down into the park's unique wetlands, symbolises Australia's failure to live up to its commitment to protect this natural heritage.

Each wet season monsoonal downpours flood the mine; the run-off carries radioactive pollutants. The water from the around the mine, called a 'restricted release zone', is directed into retention ponds where it is meant to evaporate during hot summer months. However ERA failed to design the ponds to cope with the rains deposited in unusually wet seasons. The company has lobbied persistently to be allowed to release this contaminated water into nearby Magela creek. The pollutants are being carried into the wetlands where they concentrate in the silt over the dry season. Aborigines living downstream, whose food come from hunting, fishing and gathering from the land, oppose such releases.

The Ranger Uranium Environmental Inquiry recommended the tailings be returned to the pit when mining ceases. ERA wanted the tailings to be left as a terraced hill. Inevitably as the hill eroded radioactive sands would drift into the creek. Aborigines insist they be returned to the pit. Even so pollutants will leach out infinitely faster than occurred before mining. At Roxby mine the tailings dam could one day be the highest landform in the area. Here too evaporation was supposed to remove all the water from the dam. However, water has seeped through underlying porous limestone formations to contaminate the ground waters.

Planned rehabilitation amounts to no more than a thin covering of soil and vegetation, a poor defence against desert winds for thousands of years. As time passes the tailings will erode and creep like sand dunes across the wind-swept arid plains towards the farmlands and towns to the south.

The Reactor Fuel

The enrichment and rod fabrication plants, which transform uranium into reactor fuel, create liquid, solid and gaseous wastes with low radioactivity. There have been innumerable accidents with nuclear cargoes carried between plants at the 'front end'. Though low in radioactivity the spillages have spread contamination far and wide. Volatile 'hex' (uranium hexafluoride) is especially hazardous both as a chemical and for its radioactivity. Its release in a serious accident, occurring in a populated area, can have catastrophic consequences for people's health.

The 'Back-End'

Every 12 months or so a reactor is closed down in order to replace one-third of the fuel rods (about 30 tonnes) with fresh ones. About a third of the uranium-235 in the rods is still 'unburned' but waste fission products smother the nuclear reactions just as ash does a coal fire. However, the nuclear 'ash' cannot be shaken through a grate; it can only be separated by removing the fuel rods.

Spent fuel contains 1 per cent plutonium and other transuranic elements and 3 per cent fission products. The remainder is mostly unchanged uranium-238. Spent-fuel straight from a reactor emits fierce heat and lethal radiation and must be handled by remote control. The radioactivity of 30 tonnes of spent fuel (from one year of operation) is equivalent to 5000 tonnes of radium. It is stored in water-cooled ponds on the reactor site.

Within 15 days the radioactivity drops to about one-tenth. Thereafter it remains a slowly decaying hazardous material for thousands of years.

Most nuclear power countries have now settled for one of two strategies to deal with spent fuel:

- A once-through fuel cycle where spent fuel is to be directly disposed of in canisters by permanent deep burial.
- A closed fuel cycle where uranium and plutonium are recovered by reprocessing. A high-level radioactive liquid waste is evaporated to a solid and incorporated into glass or ceramic blocks for permanent deep burial.

Reprocessing

Reprocessing was first developed for the isolation of plutonium for nuclear weapons. Commercial reprocessing was developed later to provide plutonium for breeder reactors designed to produce more plutonium again than they 'burn'. The first reprocessing plant was opened in 1963 at West Valley, New York State. Governor Rockefeller described it as a "unique operation and a symbol of imagination and foresight". Six years later the plant was too 'hot' to handle and had to close. Two commercial reprocessing plants presently service the world's reactors. One is at Sellafield in Cumbria, Britain and the other at Cap La Hague in Normandy, France.

Reprocessing begins by chopping fuel rods into pieces and dropping them into a tank of nitric acid. A solvent is added to extract the uranium and plutonium. The acid solution containing the fission products is separated off and concentrated by evaporation. This is high-level liquid waste which is stored in cooled tanks. The dissolved plutonium and uranium are separated from each in another solvent fractionation. The plutonium is stored in vaults as small specially shaped pieces to prevent a build-up of neutrons which could cause an explosion. The recovered uranium is depleted in uranium-235 and requires enrichment before use as a fuel.

Reprocessing plants are the source of about two-thirds of all the low-level liquid and gaseous wastes produced in the nuclear fuel cycle. At Sellafield millions of liters of effluent carrying plutonium and other transuranic elements and fission products are discharged daily into the Irish Sea. Plutonium has become absorbed on seabed silt which is washed on to nearby beaches making them dangerously radioactive. Ocean currents carry radioactive waste as far away as the Scandinavian coast.

Cumbrian people have more plutonium in their bones than any other people. Cesium-137 and strontium-90 in fish caught in the local waters are 200 times higher than fish taken from Atlantic waters. Local fish shops carry a notice "No Irish Sea fish sold here". The nuclear industry had planned to operate breeder reactors commercially by the 1990s. It has failed to happen and is unlikely to do so in the foreseeable future. This removes any commercial justification for reprocessing. Plutonium decays faster than uranium; if not used as fuel within three years it must be reprocessed to remove decay products. Some plutonium is used with uranium in 'mixed oxide' fuel in conventional reactors. However, plutonium from reprocessing is three times more costly than uranium.

Waste Disposal

Wastes are commonly classified according to their level of activity.

- High-level waste generates heat and has to be constantly cooled.
- Intermediate-level waste is less radioactive but still needs shielding; it does not generate enough heat to require cooling.
- Low-level waste exceeds the radioactive limits for conventional landfill sites and must be buried at special sites. It is unshielded during transport.

High-level Waste

After 40 years of nuclear power there is still confusion and uncertainty about what will happen to its wastes. Not a single kilogram of high-level waste has yet been disposed of in a permanent repository. The problems are political as well as technical. Meanwhile spent fuel continues to accumulate. Three quarters of the spent fuel so far produced is still being stored at reactor sites or in interim storage depots.

In the once through cycle spent fuel becomes the waste to be disposed of. It is envisaged that the rods will be held in cool storage up to 100 years or longer to allow heat and radiation levels to drop to manageable levels. According Swedish nuclear authorities the rods will eventually be prepared for disposal by placing them in copper canisters about a meter diameter and four meters long. Copper powder will be poured into the space between the rods. The canisters will then be placed in a deep repository. In the closed cycle the high-level liquid waste from reprocessing is solidified. France and Britain have plants for putting waste into glass. Japan and the United States plan to use ceramics.

The waste is first evaporated to a solid powder. A year's waste from a large reactor is reduced to about a cubic meter. This is a relatively small volume but it is horrendously lethal retaining radioactivity equivalent to as much as 10 tonnes of radium. Glass or ceramic powder is mixed with 10 per cent of the evaporated powder, melted and poured into canisters. The canisters are to be held in cold storage until a decision is made on what will be their final resting place.

The uncertainty surrounding the ultimate fate of the solidified waste is evident in the recommendation of the British Waste Management Advisory Committee to store the solidified waste "for at least 50 years ...a decision could then be taken on whether to continue to store these wastes: deep underground, on the ocean bed or under the ocean floor". After vigorous protest by Greenpeace in their Zodiac boats on the high seas a moratorium on ocean dumping of radioactive waste was declared in1994. The London Dumping Convention later banned ocean dumping permanently.

Permanent Deep Disposal

The solution generally favoured by nuclear industry to get rid of its high-level waste is to bury it in geological formations of granite or basalt. Kilometer deep holes would be bored into rock, canisters lowered into them and then they would be sealed off with concrete. Or else caverns or tunnels will be dug deep into rock or salt formations. Sweden, decided by referendum to cease generating nuclear power by 2020, has chosen to use the 'once through' cycle. It plans to build a repository, called a 'laboratory', consisting of a tunnel, 3500 meters long and spiralling down 500 meters.

Canisters will be deposited in holes, drilled into the rock floor, and covered with bentonite powder and sand. When the tunnel is full it will be sealed off with concrete. The copper canisters are expected to erode badly within 1400 years. By then almost all fission products will have decayed to stable elements but plutonium and other transuranics will have lost little of their activity. Effectively the only barrier against the plutonium escaping is the rock. Says Nils-Axel Morner, a leading Swedish geologist on rock movement: "Never trust a rock". That a rock has survived a million years says nothing about its future stability...When one piece is pushed the whole thing moves".

Radioactive Rubbish

Nuclear industry like other industry creates bulky rubbish paper, plastics, gloves, clothing, equipment, tools with the difference that it is radioactive. Some of this rubbish was buried at sea until action by Greenpeace and nations with fishing interests caused the practice to be banned. Much of this radioactive rubbish has been buried in shallow land graves or sometimes in disused mines. Plutonium and other radioactive elements have migrated from these dumps into the surrounding environment.

The more highly contaminated rubbish, classed as intermediate level, is made up of the cladding on fuel rods, glove boxes for handling plutonium, sludges and other waste contaminated with long-lived transuranic elements. Britain will bury intermediate-level waste in trenches roofed and lined with concrete. Waste, broken into bits, will be dumped into the trenches and mixed in concrete 'like currents in a cake'.

Politics of Waste Disposal

People naturally do not want nuclear waste on their doorstep. In 1986, the British government announced sites for dumping low-level waste. One politician from a designated area, conservative though he was, threatened that his constituents would "launch a guerilla war against the authorities". France is held up as an example of a country where people are reconciled to nuclear power. If so it does not extend to the waste. All government attempts to test suitable rock formations at selected sites have had to be abandoned in the face of vehement public protest. The French government, like most other governments, has shelved preparations for a deep repository "until sometime next century".

The Beijer Institute, in Sweden, analysed waste disposal plans of the nuclear countries. It concluded any solution to the problem of disposing of high-level nuclear waste is 'transcientific' that is beyond the bounds of science. For the safety of any waste repository, over a hundred thousand years and more, can never be proved. The data on waste performance can only be gathered over a short time span compared to the time wastes remain hazardous. In this respect nuclear wastes differ from other wastes.

Technical solutions, like the Synroc process, touted by the Australian Nuclear Science and Technology Organisation, have no relevance to the timeless hazard of the waste. Any scientist with a solution will not live long enough to test their 'discovery'. Clearly the nuclear waste problem outweighs any gain from no carbon dioxide emissions. We must look to another energy source, which also emits no carbon dioxide the renewable solar energies.

A National Repository

The federal government has begun studies on finding a suitable site 'for low-level and short-lived intermediate level radioactive waste'. Overseas practice suggests that either shallow-land burial or near-surface concrete structures will be used. The waste would be paper, plastics, glass, clothing, metals, instruments and soil. The waste comes from medical practice, research and industry. Suitable sites are being sought at Maralinga and around Cooper Pedy in South Australia, around Broken Hill in New South Wales and at Mount Wilson in Western Australia. All governments support the concept of only one national repository but all are shy about offering to host it. Only the NT Government expressed interest but after a feasibility study it too lost interest.

Clearly community opposition to radioactive waste, even when it is low-level, being dumped in their state is the reason for politicians being so shy about hosting a national repository. Conservation groups argue that responsibility should lie with the waste makers to care for their own waste. That would encourage them to minimise their waste making or find alternatives. Since the repository is to be commercial those in charge will be likely to encourage waste formation and even seek to import nuclear waste to get the highest return on their investment. During the past decade, the UK power industry has undergone a technological transformation.

Starting in the late 1980s, the phenomenon known as the 'dash for gas' has led to the construction of a large number of power stations fuelled by natural gas. These power stations are based on a new technology, the Combined Cycle Gas Turbine (CCGT), which was previously confined to the fringes of an industry fuelled by coal, oil and nuclear energy. The privatisation of the UK's electricity and gas industries changed the rules governing new investments and triggered a wholesale switch to cheaper, cleaner CCGTs. Whilst the dash for gas represents a particularly powerful manifestation of a global trend towards the use of CCGTs, it is also important because of the severe effect on the UK's deep mine coal industry. For this industry, the replacement of traditional coal-fired power plants and the failure of newer cleaner coal technologies to attract the attention of private sector investors represent the latest stage in a long process of decline.

Amongst these new technologies, the fluidised bed boiler had been regarded as the coal industry's best hope of survival. However, despite limited progress in some parts of the world, this technology has consistently failed to meet the expectations of its supporters. The aim of this chapter is to explain the success and failure of two fossil fuel power generation technologies that both produce cleaner electricity than their traditional counterparts. Whilst the reasons for the CCGT's global triumph and the failure of the fluidised bed appear to be relatively clear cut to current investors, this outcome was far from obvious just 15 years ago. The key question is how the CCGT came to be in the position to take advantage of a fortuitous combination of circumstances in a number of different countries. This question will be examined in the light of theories drawn from the evolutionary economic and social constructivist traditions. The conclusions of the chapter will examine the wider implications of success and failure for governments wishing to aid the development of more sustainable energy technologies.

Inside Two Technologies

The Combined Cycle Gas Turbine (CCGT)

The Combined Cycle Gas Turbine (or CCGT) represents a radical departure from traditional power generation technologies. By combining two established building blocks - the gas turbine and the steam turbine - CCGT designers have achieved electrical efficiencies which are almost 50% higher than those of other fossil fuel power stations. Most CCGTs work in the following way: Fuel (usually natural gas) is burned in an industrial gas turbine, to generate electricity and waste heat. The waste heat is then passed into a heat recovery boiler which uses the gas turbine's hot exhaust gases to generate steam. This steam is then used to drive a small steam turbine and produce some more electricity. In some cases, more than one gas turbine is used (each with its own steam generator) in combination with a single steam turbine. However, in all modern CCGTs the steam turbine is designed so that it produces approximately one third of the total power output, with the remaining two thirds being met from the gas turbine(s).

The most technologically advanced part of every CCGT is the industrial gas turbine. Most gas turbine designs have been developed steadily over the last 50 years from their roots in aircraft jet engines. Although the industrial gas turbines most frequently used in CCGTs are much larger than the aircraft variety, they still work on the same principle. For designers, there are a number of ways to increase the efficiency of a gas turbine and hence, the overall efficiency of a CCGT. The traditional approach is to raise the temperature at which the combustion gases enter the turbine. Achieving a temperature increase in this way is not difficult - the real challenge is to ensure that the metal blades which make up the turbine do not overheat the process. Turbine blade design is a sophisticated business requiring the use of the latest high temperature metal alloys. At the same time, advanced casting techniques are necessary to incorporate a network of tiny holes within the hottest blades for cooling air. Until recently, many of these advanced materials, casting techniques and cooling technologies have been readily available. As one seasoned observer puts it, all the gas turbine manufacturers had to do was 'go shopping at the aero-engine technology supermarket'.

This supermarket is packed with products from the multi-billion dollar military jet engine programmes of the US Department of Defense and their European counterparts. The newest generations of CCGT have begun to depart from this established trajectory as designers search for further efficiency gains. As new products such as the H System from General Electric demonstrate, the move to efficiency levels beyond 60% will require an increasing amount of innovative thinking. The Fluidised Bed Boiler shows a simplified layout of a fluidised bed power plant. This technology gets its name from the way in which the combustion process is managed. The fuel, together with a sorbent for sulphur removal, is fed into a heated bed of inert material such as sand. The bed is 'fluidised' (i.e. made to behave as if it were a liquid) by a constant stream of air, pumped in from below. As a consequence of this, combustion of the fuel is much more uniform and occurs at a lower temperature than it would in a conventional boiler. The typical bed temperature of 900C also maximises sulphur removal and minimises NO x emissions.

Although the fluidised bed process is complex, the basic differences from a conventional fossil fuel boiler are the low combustion, the fluidising air and the removal of sulphur during combustion. The latter property, together with its ability to burn a wide variety of difficult fuels (e.g. coal processing wastes), is what really sets this technology apart from its predecessors. It is characterised by relatively low air velocities which give the bed a discernible upper surface. Later varieties known as circulating fluidised beds (CFBs), use higher air velocities to propel the bed material into the upper portion of the boiler. CFB designs normally feature an additional component, known as a cyclone, which recycles unburned fuel back into the bed. The effect of this greater movement of the bed material is to enhance fuel mixing and combustion efficiency. A final development in fluidised bed technology, which has yet to be fully demonstrated commercially, is the pressurised fluidised bed (PFB).

This particular variant works in much the same way as the others. However, the boiler is now pressurised in order to produce exhaust gas at sufficient temperature and pressure to drive a small gas turbine. This gas turbine is then used to generate extra electrical power. Whereas CFBs and BFBs yield maximum efficiencies of around 38%, PFB plants offer the possibility of efficiencies of up to 42%. For all types of fluidised bed, unit sizes are limited, with maximum power outputs of around 250MW.

An Open and Shut Case

There is a stark contrast between the commercial success of the CCGT and the limited interest in fluidised bed technology. At present, 300 Gigawatts (GW) of CCGT capacity is in operation or under construction world-wide, whilst installations that use fluidised bed technology have an equivalent combined output of less than 20GW. Whilst there are large numbers of CCGTs in many countries in Europe, Asia and North America, fluidised bed plants are concentrated in a small number of countries. A simple economic comparison of these two technologies using typical UK fuel prices reveals why this is the case. This is not often the case at present since low international coal prices are accompanied by relatively cheap natural gas in many areas of the world.

Even if coal prices are sufficiently cheap, experience shows that it is far more likely for coal-fired power plants to use traditional boiler technology. The main reason for the CCGT's economic advantage is that the capital equipment is very cheap to buy in comparison with most other generating options. This characteristic is particularly appealing for the new breed of private power developers since they finance new projects through high interest bank loans rather than their own balance sheets or State funds. This built-in economic advantage is reinforced by the fact that CCGT construction times are very short. New plants can be completed in less than two years rather than the four to ten years that characterise other large scale technologies. For many private investors, this further accelerates the repayment of bank loans since their only source of revenue is the sale of electricity.

The CCGT also has important environmental advantages that are attractive for governments as well as investors. CCGT emissions are lower than those from other fossil fuel technologies. A typical CCGT emits around 65% less carbon dioxide than a traditional coal- fired power plant for each unit of electricity generated, almost no sulphur dioxide and relatively small quantities of oxides of nitrogen. In addition, CCGTs create a small visual impact, and may be sited nearer to population centres than large coal or nuclear plants. This makes it easier for developers to secure planning permission. Whilst these advantages provide clear evidence for the CCGT's current popularity, they do not explain how this technology came to be in such a dominant position. As Donald MacKenzie has argued, although hindsight may suggest that a particular technology is intrinsically superior, it is more important to determine how this state of superiority has come about. Therefore the search for an explanation of success needs to take in historical perspectives on technology development.

A New Gas Turbine Paradigm

Jorge Islas has argued that the rise of the industrial gas turbine, the technology at the heart of the CCGT, signals a paradigm shift for electric power generation technology. In doing so, he takes his lead from the evolutionary economics literature. For Giovanni Dosi, a prevailing technological paradigm 'embodies strong prescriptions on the directions of technical change to pursue and those to neglect'. The selection of this paradigm and the development of a technology within it are influenced by many economic, institutional and social factors. Jorge Islas asserts that the end of the old steam turbine paradigm is associated with a number of economic, political, environmental and technological forces that have become apparent since the first oil shock in 1973.

These include the fact that 'conventional large power stations entered a phase of saturation (slowing of technical progress, exhaustion of scale economies, diseconomies connected with growing complexity of systems of generation) and the appearance of environmental constraints and nuclear security [concerns]'. The formation of a new gas turbine paradigm has been encouraged by 'changes in the organisation of the electricity industry inspired by economic "neo-liberalism", seen as a remedy to the crisis in the electricity industry, [which] permitted the entry of new electricity producers'. Jorge Islas' approach leads us to conclude that an old paradigm has been abandoned in favour of a new one. It appears to offer a convincing explanation for the success of the CCGT, a technology at the forefront of the gas turbine paradigm, and the failure of the fluidised bed which may be associated with the old steam turbine paradigm.

Although it is certainly the case that fundamental shifts have taken place, it is questionable whether the paradigm shift has been so universal. Sales of traditional steam turbine power plants are still buoyant, particularly in countries such as India and China which have access to large coal reserves. The gas turbine paradigm has not yet had any impact on new coal- fired power plants in these countries since the technologies that enable gas turbines to burn gasified coal are still under development. Such caveats highlight two major drawbacks of the paradigm concept. Firstly, the time at which an old paradigm ends and a new one begins is not always easy to define. Although the oil shock of 1973 marked a period in which many of the technological and institutional traditions of the electricity industry were shown to have fundamental flaws, there many other points in history which could be the start of the gas turbine paradigm.

These include the start of the 'gas turbine age' in the late-1960s, the transition of the gas turbine to mass production as the jet engine after World War II, or the period in which electricity privatisation and liberalisation were implemented in practice during the 1980s. The second problem that arises with the application of paradigms relates to their scope. It is not clear whether we should be thinking in terms of a gas turbine paradigm (to succeed a steam turbine paradigm), a small power station paradigm (to succeed a large power station paradigm), or a cleaner technology paradigm (to succeed a 'dirty' technology paradigm).

Each of these cases partly characterise shifts which have already happened or those which may take place in the future. However, the CCGT and the fluidised bed do not fit precisely into the new and the old paradigms respectively. Therefore, as a tool to explain success and failure for this particular case, paradigms are of limited use. Of more relevance is Jorge Samperio's subsequent view that the gas turbine has reversed the lock-in of the steam turbine. This analysis allows for the fact that steam turbine technology is still being used in a large number of new power plants, including CCGTs.

The Interpretative Flexibility of the Gas Turbine

Apart from the ambiguous nature of the paradigm, the main limitation of evolutionary economics is its narrow focus. Whilst there is some consideration of 'the role often played in the establishment of a particular technological trajectory by public ("political") forces', the primary concern is still economics. Political forces are essentially seen as an adjunct to the theory which is not explored in great depth, except with reference to military and space projects. As a result, evolutionary economics does not explicitly provide a framework which allows the influence of many different actors (primarily governments, but also consultants, utilities, financiers, manufacturers etc.) on success and failure to be clearly analysed. It is therefore necessary to draw on some concepts which stem from sociological theories of technology development, particularly the Social Construction of Technology (SCOT) approach.

The most important characteristic of this approach is the lack of rigid boundaries between the economic, the social, the political and the technological. It emphasises the need for a stable socio-technical network of interrelated actors around a new technology in order for it to succeed. As Thomas Hughes has argued, there is a strong rationale for a network-based analysis of the electricity industry that takes into account economic, social and political factors: Technological systems contain messy, complex, problem-solving components. They are both socially constructed and society shaping. Among the components in technological systems are physical artefacts, such as the turbo generators, transformers, and transmission lines in electric light and power systems. Technological systems also include organizations, such as manufacturing firms, utility companies, investment banks, and they incorporate components usually labelled scientific, such as books, articles, and university teaching and research programmes.

Legislative artefacts, such as regulatory laws, can also be part of technological systems. Because they are socially constructed and adapted in order to function in systems, natural resources, such as coal mines, also qualify as system artefacts. For this chapter, the most important SCOT concept is interpretative flexibility, a term which explains how different groups of actors or relevant social groups attach different meanings and problems to the same technology. For example, Wiebe Bijker has observed that during the development of the bicycle, 'the high- wheeled Ordinary [which became the Penny-farthing] was at once a dangerous machine, prone to failure in the marketplace, and a well-working machine that allowed highly skilled physical exercise, resulting in a commercial success'.

The implication is that if a technology is particularly flexible, it could support these different meanings simultaneously. As this chapter will show, key relevant social groups for the CCGT include the power plant equipment industry, the aircraft engine manufacturers and electric utilities. However, generic gas turbine technology stands out with its particularly high degree of interpretative flexibility. To the relevant social group of the armed forces, the gas turbine is the most efficient propulsion unit for fighter aircraft as well as Naval frigates. The relevant social group of electric utilities sees the gas turbine as the most important component of their current technology of choice, the CCGT. Elsewhere, the relevant social group of oil and gas companies uses this technology for pumping gas through long pipelines and to provide power for offshore installations. In addition, the gas turbine is also a large source of revenue and a strategic core technology for the relevant social groups of power plant equipment manufacturers and aircraft engine producers.

The Struggle for Acceptance

During the past 50 years, the interpretative flexibility of the gas turbine has allowed the steady and successful enlistment of the electric utilities as a relevant social group. The first attempt to sell the industrial gas turbine to the electric utilities began as early as the late 1940s. However, this attempt failed because this new technology was outperformed by steam turbine power plants which were going through a phase of rapid technical change. As a result, the gas turbine was too small and inefficient for most electric utilities. A typical industrial gas turbine, with an output of 5MW and a thermal efficiency of up to 20%, was unable to match fossil-fuel steam plant sizes of up to 100MW and thermal efficiencies of at least 25%. The gas turbine equipment manufacturers responded by developing complex variants of their designs, with theoretical efficiencies of up to 34%.

However, the few machines that were built suffered from acute reliability problems. For the fluidised bed boiler, the prospects of utility interest were even more remote at this stage since the idea of using this technology for steam generation had not yet emerged. As a result, the links between both technologies and the relevant social group of electric utilities remained slight at best. Despite the lack of utility interest in the industrial gas turbine of the 1950s, generic gas turbine technology was not short of supporters. Following the reliability problems of complex gas turbines, the return to simple designs by the equipment manufacturers led to the enlistment of the oil and gas industry. For this relevant social group, the gas turbine was not primarily a device to generate electricity. Instead, its interpretative flexibility meant that it was an economical means of pumping natural gas and oil through long distance pipelines.

By the end of the 1950s, the limited success enjoyed by the industrial gas turbine had encouraged the equipment manufacturing companies to experiment with their first CCGTs. However, the CCGT remained peripheral to their power generation business for another thirty years. By contrast, the airborne counterpart of the industrial gas turbine, the jet engine, had already emerged as the dominant technology for aircraft propulsion. The Korean War had provided the incentive for the expansion of military R&D spending to levels which would be sustained for the next forty years. In addition, the first jet powered civilian airliners had begun to enter service in Europe and North America. The support of this network of relevant social groups allowed generic gas turbine technology to flourish in a way that would eventually aid the enlistment of the electric utilities by the CCGT.

Enlisting the Electric Utilities

The chain of events that eventually led to the first substantial link between the electric utilities and the industrial gas turbine may be traced to the power blackouts in the UK and North America during the early and mid-1960s. These blackouts exposed the lack of provisions to cope with such emergencies within the electricity supply grids of most industrialised countries. In the years that followed, the utilities in these countries installed a large number of emergency gas turbines to restore electricity supplies. The fast start-up capability of both the industrial and aircraft engine-derived variants of gas turbine meant that they were ideal for this type of application. Although hindsight provides evidence that the 1960s blackouts could have been predicted and avoided, they can be viewed as a classic unpredictable historical event.

The subsequent increase in demand for industrial gas turbines allowed the equipment manufacturers to re-invest their revenues and improve their designs. These improvements, which yielded increases in unit size and efficiency, provided the utilities with increasing returns from their adoption of the gas turbine. It is this process which eventually led to the emergence of the CCGT. The most important consequence of the 1960s power blackouts is the establishment of a vital lead for gas turbine technology over the fluidised bed for power generation applications. Although the idea of using fluidised bed boilers for steam generation (and, by implication, power generation) emerged during this decade, the amount of practical development work remained small until the mid-1970s. By this time, the first CCGTs had been constructed in the USA. Once the electric utilities had become more familiar with their new gas turbines, it was much easier for the equipment manufacturers to incorporate larger versions of these machines into CCGT plants. Indeed, some of the American equipment manufacturers reported that the electric utilities began to enquire about gas turbines for continuous rather than emergency service.

At this point, the interpretative flexibility of this technology was vital since it allowed the manufacturers to import advanced materials and cooling techniques from the jet engine in order to scale-up their designs to a useful size.

The advent of gas turbines with capacities of over 50MW allowed the construction of 100MW CCGT blocks which operated with a thermal efficiency of around 40%. Several of these blocks could be constructed side by side in order to form a medium-sized power station that matched the efficiency of contemporary steam turbine power plants. This combination of factors led to the enlistment of the first batch of electric utilities by the supporters of the CCGT.

Many of the first orders were placed in the USA, where a bubble of cheap natural gas had improved the economics of gas-fired power. In addition, General Electric (GE) and Westinghouse were technologically better placed than the European manufacturers as a result of their current and previous aircraft engine connections. At the same time, the fluidised bed boiler finally became attached to the same socio-technical network with the construction of the first demonstration plants for electricity generation in the UK and the USA.

Temporary Interest in the Fluidised Bed Boiler

The new wave of interest in the CCGT as a power generation option did not last for long. Its end, like its beginning, was precipitated by an unpredictable historical event. This time, it was the decision of the Middle-Eastern OPEC nations to quadruple their oil prices which had a large impact on the CCGT since it was accompanied by a sudden increase in natural gas prices. The problems of the CCGT were subsequently compounded by a number of serious reliability problems which affected the newly scaled-up industrial gas turbines. For the fluidised bed boiler, the 1973 oil shock represented a much needed boost to its prospects. An attitude of indifference amongst relevant social groups such as governments, utilities and large boiler makers was replaced by a growing interest. Faced with the twin problems of high oil prices and localised environmental pollution, governments and equipment manufacturers in North America, Europe and Japan began to look at ways of burning coal more cleanly and efficiently.

Since fluidised bed technology offered the prospect of fulfilling the first (and perhaps the second) of these criteria, a number of demonstration projects were initiated. The fluidised bed was not the only beneficiary of the new public R&D programmes. These programmes also included ambitious plans to develop high efficiency industrial gas turbines and jet engines, some of which eventually yielded CCGT improvements. By the late-1970s, utilities had formed an attachment to both of our two technologies, though the extent of this attachment was still weak. Although the demand for gas-fired power had almost disappeared outside the Middle East, the CCGT's position remained much stronger than that of the fluidised bed. The existence of a significant number of plants in commercial service in the USA and Europe, together with the strong network of relevant social groups that were continuing to support the jet engine, placed this technology in a strong position for the future. By contrast, the fluidised bed boiler was set to remain a demonstration technology for several more years despite the support of some governments, large boiler makers and utilities.

Although the first oil shock appeared to signal an abrupt end to the new gas turbine age, the development of the CCGT continued throughout the 1970s and 1980s. Trends in capital costs, environmental performance and reliability combined to increase the advantages of the CCGT over the alternatives. Whilst the performance of traditional technologies did not improve substantially, the CCGT of the late 1980s was substantially more attractive than its predecessor of the early 1970s. Even though orders for gas turbines and CCGTs remained scarce until the end of this period, the CCGT was well placed by the time that oil prices fell in 1986. Faced with such rapid improvements the fluidised bed boiler was left behind, even though its supporters made considerable progress in terms of unit size and reliability.

The figures, which are all taken from estimates under conditions in the USA, have been adjusted to year 2000 prices. Since 1980, there has been a substantial increase in the capital costs of conventional fossil- fired steam plants whilst the cost of a typical CCGT has actually fallen slightly. According to some estimates, the cost of a typical large coal-fired station has escalated by 20% as a result of the need for flue gas scrubbers and cooling towers (Joskow and Rose). Although they appear to be large, these cost escalations are not as substantial as those for nuclear power stations. The reasons for the downward trend in CCGT costs may be traced to the intense competition for orders within the heavy electrical industry as well as rapid advances in thermal efficiency.

Although historical capital cost data is difficult to locate, it is possible to conclude that the fluidised bed failed to outperform traditional fossil-fired steam plants in economic terms. The early fluidised beds that were constructed in the USA following the introduction of the Public Utility Regulatory Policy Act of 1978 featured an economic advantage over a traditional coal- fired plant with flue gas desulphurisation to reduce sulphur dioxide emissions. It was at this time that interest rose in this technology as a way of burning coal and other polluting fuels whilst minimising emissions. However, this advantage was soon lost as the proponents of traditional boiler technology fought back.

As well as increasing the comparative economic advantage of the gas turbine and CCGT, the equipment manufacturers also had to restore utility confidence in the reliability of their designs. Attitudes during the mid-1970s were typified by one utility executive who is reported to have said that 'the only good gas turbine is one that doesn't have to run'. If the relevant social group of electric utilities were going to be encouraged to renew and strengthen their attachment to the CCGT once the fuel price situation changed, the image of this new technology would have to be improved. Despite this dismissive attitude from some quarters, the utilities played a crucial role in the improvement of reliability, particularly through the efforts of the Electric Power Research Institute in the United States.

When oil and gas prices fell in 1986, their reluctance to embrace CCGT technology did not last long because the steady efficiency improvements implemented by manufacturers began to yield substantial economic benefits. These benefits have outweighed concerns about reliability even though some newer CCGT designs have suffered from significant technical problems. Asian utilities such as the Tokyo Electric Power Company and the Electricity Generating Authority of Thailand were the first to recognise the CCGT's potential. They were accompanied by the new breed of independent power companies in the UK and the USA, who were keen to use the CCGT as a way of entering newly liberalised electricity markets.

Powerful Relevant Social Groups

Whilst the evidence presented so far shows how the CCGT eventually enlisted electric utilities and independent power companies through its interpretative flexibility, an exclusive focus on these relevant social groups misses out some important ingredients of success. This chapter has already alluded to the influence of other relevant social groups on success and failure, namely the power plant equipment manufacturers and various government agencies. Before bringing the analysis to its conclusion, it is important to acknowledge the power of these relevant social groups to shape success and failure. SCOT approaches to technical change have already addressed the issue of power in previous empirical case studies. In his social construction of the fluorescent light bulb, Wiebe Bijker introduced power 'to account for the obvious differences in economic power between some of the relevant social groups ... including Mazda companies a lighting cartel, utilities, independents, consumers, fixture manufacturers, and the government'.

The International Equipment Companies

The world's dominant manufacturers of equipment for electric power plants have been in existence for over 100 years. Companies such as Siemens of Germany and GE of the USA were founded by electricity industry pioneers during the late 19th century. For at least the first sixty years of its existence, the heavy electrical industry operated through a series of cartels. These included the famous Phoebus electric lamp cartel of the 1920s and the International Electrical Association which acted as a cartel for at least 40 years from the mid-1930s. Illegal activities such as price fixing and market allocation were common within these cartels, as were more legitimate associations through product licensing agreements. As a result, the heavy electrical industry has always been a financially strong and powerful relevant social group for the technologies it championed.

One of the defining features of the dominant companies within heavy electrical industry is their strong technical competencies in the area of turbine design. Companies such as GE, Westinghouse and Siemens were the first to take up the steam turbine following its invention by Charles Parsons in 1884. They were also heavily involved in the jet engine development programmes during and after World War II, and were the first companies to develop the land based gas turbine as a new electricity generation technology. The turbine- based competence of these companies meant that they were able to subsequently develop and sustain an interest in CCGT design and manufacture. The success of the dominant heavy electrical industry firms in commercialising turbine-based technologies such as the CCGT contrasts sharply with their general lack of interest or skills in boiler equipment manufacture. As one manager from GE explained, his company felt that there 'was very little unique contributed value in putting together a bunch of tubes and putting a shell around it'.

In fact, the first efforts of gas turbine manufacturers such as GE and Westinghouse to build heat recovery boilers for their CCGTs were a disaster, and they soon contracted out this activity. Similarly, their performance in the construction of nuclear steam generators was much weaker than rival boiler makers like Babcock and Wilcox and Combustion Engineering. The upshot of this technical specialisation is that the largest power plant equipment manufacturers only showed limited interest in new boiler-based technologies such as the fluidised bed. Instead, this technology was developed on the sidelines by a number of small and medium-sized boiler makers, and was only picked up by larger equipment companies as a by-product of industry take-overs during the 1980s and 1990s.

Whilst the CCGT emerged as the fastest developing power plant technology with a global market share approaching 50% (Financial Times, 2000), the boiler makers became less and less technologically important. Although orders for power station boilers have continued, the CCGT boom has strengthened the position of the big four heavy electrical industry companies (GE, Siemens-Westinghouse, Alstom and Mitsubishi) at the expense of the boiler makers.

A Political Economy of Technology Choice

Governments and their agencies influence the success and failure of energy technologies in many different ways, some direct and some indirect. In doing so, governments can comprise several relevant social groups which become linked to a particular technology for a variety of different reasons. The UK dash for gas provides a good illustration of some of these links. As stated earlier, the dash for gas led to the rapid diffusion of CCGT technology within the UK electricity industry starting in the early 1990s. This phenomenon is particularly relevant here since it confirms Wiebe Bijker's observation that the 'stabilisation of artefacts is a social process and hence subject to choices, interests, and value judgements - in short, to politics'. In the early 1990s, parts of the UK government acted as particularly powerful relevant social groups for the CCGT. Their clearest influence can be traced to the privatisation and liberalisation of the electricity supply industry which began in 1989.

Strong government support for privatisation and the development of competition led to the encouragement of independent power producers to challenge privatised generating companies. As this chapter has already illustrated, the post-privatisation financial climate gave strong incentives for power companies to choose CCGT technology for new power plants. However, obstacles remained to new entry based on CCGTs due to legislative restrictions on the use of natural gas as a power station fuel. The government Minister responsible, Cecil Parkinson, chose to overrule both UK and European laws to allow the first batch of CCGT stations to be built. As a fortuitous by-product of the CCGT's popularity, the government was subsequently able to meet its international commitments to curb emissions of carbon dioxide and sulphur dioxide without the need for further intervention.

Alongside the government's economic and environmental reasons for encouraging the dash for gas, there was a deeper political motivation for its actions. Following their return to power in 1979, a key part of the Conservative Party's agenda was the desire to curb trade union power in the aftermath of the 1978-79 'winter of discontent'. Since the UK's coal mining union (the National Union of Mineworkers or NUM) represented the more militant strand of trade unionism, it became the primary target of government efforts. Even though the power of the NUM was severely diminished following the strike of 1984-85, the dash for gas may be seen as a convenient way to weaken its influence in the UK energy system.

This additional ideological driver helps to explain why the shift in favour of the CCGT in the UK has been much more pronounced in the UK than in other countries. It also partly accounts for the continued absence of the fluidised bed and other cleaner coal technologies from the UK power generation scene. In other industrialised coal producing countries such as Germany and the USA, the switch to gas has not been so rapid and the fluidised bed boiler has made a limited impact. Whilst the UK government relevant social groups were not dominated by people who were positive advocates of the CCGT, they became firmly linked to this technology by default.

To use the terminology of Bruno Latour, the UK government was translated to support the CCGT due to the combined effect of the Conservative Party's agenda of economic liberalisation and its crusade against trade union power. In the light of these considerations, it is not surprising that the introduction of competition into the electricity was carried out in a way that was extremely detrimental to the coal industry. In addition, it is now clear why the UK dash for gas was allowed to continue unabated until the Conservatives left office in 1997.

Technology and Sustainability

This chapter has shown why one power generation technology has become a world-wide success, whilst another equally promising technology has failed to make an impact. Although the reasons for the CCGT's popularity are clear under current economic conditions, it is less apparent how this technology came to be in such a position of strength. The explanation for its ascendancy involves a complex array of factors. It has been shown that the economic dimension of the CCGT's development provides insufficient evidence for its success. Substantial weight must also be given to the structure of the electricity and power plant equipment industries, the role of unpredictable events and political motivations particularly within the UK government. At the heart of the explanation of success and failure lies the interpretative flexibility of the gas turbine, the most important CCGT component.

This flexibility has allowed the gas turbine to develop rapidly, first as the aircraft jet engine and subsequently as a power source for the oil, gas and electricity industries.

Through the application of jet engine technology to land-based gas turbines and improvements in reliability, the relevant social group of electric utilities was enlisted as a strong supporter of the CCGT in the late 1980s and early 1990s. Whilst some fluidised bed boilers have been built as a result of stringent environmental regulations in a few countries, its position has been undermined by the arrival of cheap natural gas. In conclusion, it is important to examine the implications of success and failure for future government policies, particularly those for the encouragement of more sustainable energy technologies.

The experience of government support for new energy technologies suggests that the most direct policy options - particularly R&D funding – do not necessarily lead to commercial success. In face, much of the evidence points to the opposite conclusion - that the success or failure of a given technology has little to do with government R&D funding of it. Examples of failed State-funded technologies include the fast breeder nuclear reactor and the supersonic passenger aircraft. The axiom that government funding and success have little in common would appear to be supported by the cases of the CCGT and fluidised bed. Whilst the fluidised bed boiler has benefited from considerable public financial support, most notably from the US government, it has not become a mainstream technological option. By contrast, the conventional wisdom with regard to the CCGT is that it was developed by private companies in response to demand for a cheap, environmentally friendly and trouble free source of electrical power.

Although this particular contrast between the CCGT and the fluidised bed boiler reflects some of the truth, this chapter has revealed a more complex picture. The close link between the gas turbine and the aircraft jet engine means that the current generation of CCGTs have benefited from billions of dollars of public R&D support. As a result of the military jet engine connection, the armed forces and defence agencies of many countries comprise a particularly powerful relevant social group for the gas turbine and, by implication, for the CCGT. Similarly, some observers are of the opinion that the most successful designs of fluidised bed were developed by private companies with a minor role played by government programmes. However, the fluidised bed has not benefited from obvious State-sponsored spin-offs due to its lower degree of interpretative flexibility. The implications for governments wishing to support the development and use of more sustainable energy technologies are complex.

The cases of the CCGT and fluidised bed do not give any encouragement to those who advocate 'do nothing' governments - governments which provide minimal direct support for the development of new technologies and those which prefer to leave such matters entirely to the market. It has been shown that it is highly unlikely that the CCGT would have been a success if it had not been helped by billions of dollars of government money through military budgets. It has also been established that the fluidised bed boiler has failed despite considerable public financial commitments, particularly from the US government. Such unpredictable effects reinforce the conclusions drawn by previous studies that a government strategy that supports a diverse range of technologies is likely to be the most effective.

Along with other evidence presented in this chapter, these effects also add weight to the view that technologies are not developed from concepts to commercial successes in a linear or path dependent way. Rather, they are the result of complex interactions between various relevant social groups which have to be enlisted as part of a new, stable socio-technical network. Even though new technologies may seem to be practically feasible as well as more sustainable, there are often large barriers to their adoption. For example, the relevant social group of electric utilities may not be interested in solar cells because they are too small, too expensive or too novel. Alternatively, the power plant equipment industry - a relevant social group that dominates the power generation equipment market - may be impossible to enlist because they do not have the competencies or the inclination to adopt a given technology.

The fluidised bed boiler is a case in point. In view of these potential barriers, it is not enough to provide lists of technological possibilities with an exhortation to use them simply because they are more sustainable as von Wiezsacker et al (1997) have done. Instead, mechanisms have to be devised which allow confidence to be built amongst the various parties and the establishment of new socio-technical networks, whilst allowing for the fact that some technologies will not be taken up and others will be the subject of unexpected interest. A partly successful example of such a mechanism is the UK's Non Fossil Fuel Obligation subsidy for renewable energy technologies.

Finally, from the perspective of the UK, it is easy to hold the view that revolutionary structural change is always necessary to alter established technological habits. Since the abandonment of traditional coal-fired power plants in favour of the CCGT has also led to dramatic reductions in the UK electricity industry's environmental impact, it is equally easy to conclude that such an upheaval is a pre-requisite for a shift towards a more sustainable energy system. However, the evidence from outside the UK contradicts this view.

For example, the large Japanese utilities have adopted the CCGT even though they are private monopolies. They have also reduced the environmental impact of their existing fossil-fuel power plants through an extensive programme of flue gas desulphurisation and denitrification. Elsewhere, the State-owned monopoly utilities of South Korea, Indonesia and Thailand have also built large numbers of CCGTs. Therefore a radical change in the structure of the electricity industry is not a necessary condition for the success of the CCGT. If this conclusion were generalised to other technologies, it would mean that more sustainable options can thrive within existing institutional structures.

Whilst it is may not be necessary to privatise (or even liberalise) the energy industries to encourage sustainability, some of the technologies to help combat global climate change could ultimately originate outside the established socio-technical network. Unless all of the power equipment companies follow the example of the Swiss Swedish company, ABB, and jettison their interests in traditional energy options, the progress of many new technologies may be very slow. It would appear that there is still a lot to be learned from the actions of the US government 55 years ago. Instead of giving their newly acquired jet engine to the existing piston aircraft engine suppliers, the US Airforce decided to involve the steam turbine manufacturers - manufacturers which did not have a vested interest in blocking the progress of this revolutionary new technology.

Bibliography

Andrews, B., Brewin, C. R., Rose, S., et al. 2000. "Predicting PTSD Symptoms in Victims of Violent Crime: The Role of Shame, Anger, and Childhood Abuse." Journal of abnormal psychology 109: 69-73.

Baldinger, B. G., and Nelson, D. T. 1995. "Crime Victims and Psychological Injuries." Trial 31: 56.

Bazemore, G. 2000. "Crime victims, restorative justice and the juvenile court: exploring victim needs and involvement in the response to youth crime." Violence and Abuse Abstracts 6.

Berns, Nancy S., Framing the Victim: Domestic Violence Media and Social Problems (Social Problems and Social Issues). Aldine Transaction: 2004.

Bisson, J. I., and Shepherd, J. P. 1995. "Psychological reactions of victims of violent crime." The British journal of psychiatry 167: 718.

Bjerregaard, B. 2000. "An Empirical Study of Stalking Victimization." Violence and Victims 15: 389-406.

Cling, B. J., Sexualized Violence against Women and Children: A Psychology and Law Perspective. The Guilford Press: 2004.

Cretney, A., and Davis, G. 1997. "Prosecuting Domestic Assault: Victims Failing Courts, or Courts Failing Victims?" The Howard Journal Of Criminal Justice 36: 146.

Davis, K. E., and Frieze, I. H. 2000. "Research on Stalking: What Do We Know and Where Do We Go?" Violence and Victims 15: 473-487.

Denkers, A. J .M., and Winkel., F. W. 1998. "Crime victims well being and fear in a prospective and longitudinal study." Sage Urban Studies Abstracts 26.

Dwyer, D. C. 1995. "Response to the Victims of Domestic Violence: Analysis and Implications of the British Experience." Crime and Delinquency 41: 527-541.

Epstein, J. N., Saunders, B. E., and Kilpatrick, D. G. 1997. "Predicting PTSD in Women with a History of Childhood Rape." Journal of Traumatic Stress 10: 573-588.

Erez, E., Laster, K. 1999. "Neutralizing Victim Reform: Legal Professionals' Perspectives on Victims and Impact Statements." Crime & Delinquency 45: 530-553.

Fenwick, H. 1996. "The rights of victims of crime: Protection from harassment and intimidation?" The Journal of criminal law 60: 84-93.

Gartner, R., Dawson, M., and Crawford, M. 1998/1999. "Woman Killing: Intimate Femicide in Ontario, 19741994." Resources for Feminist Research 26: 151-173.

Gaumer, C. P. 1995. "Protecting the Innocent: Victim-Witness Rights in Illinois." Illinois bar journal 83: 568.

Grant, J. M., Folger, W. A., and Hornak, N. J. 1995. "College Students' Perception of Victim Responsibility in an Acquaintance Rape Situation." College Student Journal 29: 532-535.

Handsley, E. 1997. "Suits Against the Police for Failure to Protect Victims of Violent Crime: A Feminist Perspective on the Use of Dichotomies." The Anglo-American law review 26: 37.

Henley, N. M., Miller, M., and Beazley, J. A. 1995. "Syntax, Semantics, and Sexual Violence: Agency and the Passive Voice." Journal of Language and Social Psychology 14: 60-84.

Johnson, K. W. 1997. "Professional Help and Crime Victims." The Social service review 71: 89-109.

Kenney, J. S. 2002. "Victims of Crime and Labeling Theory: A Parallel Process?" Deviant behavior 23: 235-266.

Kury , H. Ferdinand., T. 1998. "The Victim's Experience and Fear of Crime." Sage Urban Studies Abstracts 26.

Loring. M. T., and Beaudoin, P. 2000. "Battered Women as Coer[illegible]d Victim-Perpetrators." Journal of Emotional Abu[illegible]. 3-14.

Loue, S. 2000. "Intimate Partner Violence: Bridging the Gap between Law and Science." Journal of Legal Medicine 21: 1-34.

Lowenstein, L. F. 2000. "Victims of Violent Crime Leading to Post-Traumatic Stress Disorders." The Police journal 73: 256-262.

Madriz, E. I. 1997. "Images of Criminals and Victims: A Study on Women's Fear and Social Control." Gender & Society 11: 342-356.

Marietta, J. D., and Rowe, G. S. 1999. "Violent Crime, Victims, and Society in Pennsylvania, 1682-1800." Pennsylvania history 66.

Marion, N. E. 1995. "The Federal Response to Crime Victims, 1960-1992." Journal of interpersonal violence 10: 419-436.

Martens, P. L. 2002. "Immigrants as victims of crime." Violence & Abuse Abstracts 8: 3-84.

Merlo, Alida V., and Joycelyn M. Pollock. Women, Law, and Social Control. 2nd edition. Allyn & Bacon: 2005.

Nettleton, H., Walklate, S., and Williams, B. 1997. "Three Models of Probation Involvement with Victims of Crime." The Liverpool law review 19: 203.

Pain, R. H. 1995. "Elderly Women and Fear of Violent Crime: The Least Likely Victims? A Reconsideration of the Extent and Nature of Risk." British Journal of Criminology 35: 584-598.

Peltzer, K. 2000. "Risk for traumatization among violent crime victims in an urban community sample in South Africa." Curationis 23: 22-27.

Pino, N. W., and Meier, R. F. 1999. "Gender Differences in Rape Reporting." Sex Roles 40: 979-990.

Presser, L., and Gaarder, E. 2000. "Can Restorative Justice Reduce Battering? Some Preliminary Considerations." Social Justice 27: 175-195.

Ruback, R. B., Menard, K. S., Outlaw, M. C., et al. 1999. "Normative Advice to Campus Crime Victims: Effects of Gender, Age, and Alcohol." Violence and victims 14: 381-396.

Spinellis, D. 1997. "Victims of Crime and the Criminal Process." Israel Law Review 31:

Stevenson, K. 2000. "Unequivocal Victims: The Historical Roots of the Mystification of the Female Complainant in Rape Cases." Feminist Legal Studies 8: 343-366.

Tobolowsky, P. M. 1999. "Victim Participation in the Criminal Justice Process: Fifteen Years After the President's Task Force on Victims of Crime." New England journal on criminal and civil confinement 25: 21.

Varelas, N., and Foley, L. A. 1998. "Blacks' and Whites' Perceptions of Interracial and Intraracial Date Rape." Journal of Social Psychology 138: 392-400.

Weinrath, M., and Gartrell, J. 1996. "Victimization and Fear of Crime." Violence and Victims 11: 187-197.

Weisman, C. S. 1995. "Armed by Fear: Self-Defense Handguns and Women's Health." Women's Health Issues 5: 3-7.

Yee, J. L., Greenberg, M. S., and Beach, S. R. 1998. "Attitudes toward Various Modes of Coping with Criminal Victimization: The Effects of Gender and Type of Crime." Journal of Social and Clinical Psychology 17: 273-294.

Index

F

G

H

I

J

L

M

N

O

P